Geo-Environmental Problems Caused by Underground Construction

Geo-Environmental Problems Caused by Underground Construction

Editor

Bing Bai

MDPI

Basel • Beijing • Wuhan • Barcelona • Belgrade • Novi Sad • Cluj • Manchester

Editor
Bing Bai
School of Civil Engineering
Beijing Jiaotong University
Beijing
China

Editorial Office
MDPI
St. Alban-Anlage 66
4052 Basel, Switzerland

This is a reprint of articles from the Special Issue published online in the open access journal *Applied Sciences* (ISSN 2076-3417) (available at: www.mdpi.com/journal/applsci/special_issues/geo_environmental).

For citation purposes, cite each article independently as indicated on the article page online and as indicated below:

Lastname, A.A.; Lastname, B.B. Article Title. *Journal Name* **Year**, *Volume Number*, Page Range.

ISBN 978-3-7258-0042-1 (Hbk)
ISBN 978-3-7258-0041-4 (PDF)
doi.org/10.3390/books978-3-7258-0041-4

Contents

About the Editor . . . . **vii**

Preface . . . . **ix**

Bing Bai
Special Issue on the Geo-Environmental Problems Caused by Underground Construction
Reprinted from: *Appl. Sci.* **2023**, *13*, 10754, doi:10.3390/app131910754 . . . . **1**

Yanmei Ruan, Xu Luo, Jin Li, Yang Li, Shan Lin and Chengkun Ling et al.
Deformation Law of Tunnels Using Double-Sidewall Guide Pit Method under Different Excavation Sequences
Reprinted from: *Appl. Sci.* **2023**, *13*, 12764, doi:10.3390/app132312764 . . . . **3**

Guobo Wang, Tao Meng, Gan Feng, Lifeng Ma, Weimin Yang and Yi He et al.
Influence of Bedding Strength and Angle on Fracture Characteristics of Sandstone under Three-Point Bending Conditions †
Reprinted from: *Appl. Sci.* **2023**, *13*, 8216, doi:10.3390/app13148216 . . . . **22**

José-Manuel Baraibar, Miguel Gil and Iñigo Escobal
Influence of the Construction of the Urdinbide Road Tunnel on the Autzagane Aquifer in Biscay (Spain)
Reprinted from: *Appl. Sci.* **2023**, *13*, 7034, doi:10.3390/app13127034 . . . . **42**

Hai Wang, Yan Qin, Yuxi Guo and Nengxiong Xu
A Numerical Simulation of the Subsidence Reduction Effect of Different Grouting Schemes in Multi-Coal Seam Goafs
Reprinted from: *Appl. Sci.* **2023**, *13*, 5522, doi:10.3390/app13095522 . . . . **55**

Tong Jiang, Xun Pei, Wenxue Wang, Longfei Li and Shihao Guo
Effects of the Excavation of a Hydraulic Tunnel on Groundwater at the Wuyue Pumped Storage Power Station
Reprinted from: *Appl. Sci.* **2023**, *13*, 5196, doi:10.3390/app13085196 . . . . **73**

Zhenhua Zhao, Zhenjiang Luo, Hongjie Sun, Haitao Li, Qiang Liu and Haiyan Liu
Capillary Rise in Layered Soils
Reprinted from: *Appl. Sci.* **2023**, *13*, 3374, doi:10.3390/app13063374 . . . . **94**

Jun Zhang, Weili Li and Shuaihua Ye
Stability Analysis of Filled-Slope Reinforced by Frame with Prestressed Anchor-Plates under Static Action
Reprinted from: *Appl. Sci.* **2023**, *13*, 1615, doi:10.3390/app13031615 . . . . **102**

Tiantao Su, Yong Zhou, Zhengzhen Wang and Shuaihua Ye
Large Scale Model Test Study of Foundation Pit Supported by Pile Anchors
Reprinted from: *Appl. Sci.* **2022**, *12*, 9792, doi:10.3390/app12199792 . . . . **125**

Xiaobing Zhang, Xin Zhang and Shuaihua Ye
Calculation for Permanent Displacement of Single Slip Surface of Multi-Stage Loess Slope Based on Energy Method
Reprinted from: *Appl. Sci.* **2022**, *12*, 8426, doi:10.3390/app12178426 . . . . **142**

Xuwei Zhao, Keguo Sun, Yingzhou Zhen, Yiqin Hong and Huichao Zhou
Study on Share Rate of Support Structure for Super-Large Span Twin Tunnels with Small Interval
Reprinted from: *Appl. Sci.* **2022**, *12*, 7498, doi:10.3390/app12157498 . . . . **161**

About the Editor

Bing Bai

Bing Bai, male, was born on October 1966. Bing Bai is currently a professor working at the Beijing Jiaotong University, Beijing, China. He holds a Ph.D. in Geotechnial Engineering from Wuhan University, China. His research interests include geo-environmental engineering, thermal consolidation theory, and contaminant transport theory and control methods. Recently, he has devoted himself to research on advances in soil pollution and geotechnical environments. He developed a theory describing the cotransport of heavy metals and suspended particles at different temperatures in porous media and proposed a nonlinear attachment–detachment model with hysteresis suitable for substances with sizes ranging from ions to large particles, which is of great significance in understanding groundwater pollution mechanisms and in the development of purification technology. He has published more than 170 academic papers and five scientific books. He has been named in Stanford University's career/Singleyr's list of the "World's Top 2% Scientists" for three consecutive years and was awarded the Natural Science Prize of the Ministry of Education of the People's Republic of China in 2022 for his work as a first-completion recipient. He serves as an editorial board member of the *Journal of Geotechnical Engineering and Rock and Soil Mechanics* of China. He is the member of several professional committees, such as the Special Committee on Soil Constitutive Relationship, the Strength of China Society of Civil Engineering, and the Special Committee on Energy Underground Structure and Engineering of the Chinese Society of Rock Mechanics.

Preface

A geo-environmental disturbance refers to the significant changes in physical, mechanical and even chemical properties of soils closely related to the inter-related multi-physical field coupling system of solid particles, water and gas in a shallow stratum, which is caused by underground engineering construction. The evaluation of the possible geo-environmental hazards is an important topic worthy of the attention of relevant researchers and engineers in the construction of large-scale underground engineering structures. This is related to the suitability of civil engineering construction, especially for some complicated geological conditions such as saturated super-soft soil, laterite, loess, sand gravel stratum, saline soil, frozen soil and karst stratum with rich water.

This Special Issue aimed to present the recent developments in the interaction between the coupled multi-physical fields and underground structures, as well as geo-environmental effects, to stimulate fruitful technical and scientific interactions between professionals.

A total of ten papers regarding geo-environmental problems in various fields caused by underground construction are presented in this Special Issue Reprint.

Thanks to all the authors and peer reviewers for their valuable contributions to this Special Issue, "Geo-Environmental Problems Caused by Underground Construction". I would also like to express my gratitude to all the staff and people involved in the creation of this Special Issue Reprint.

Bing Bai
Editor

applied sciences

Editorial

Special Issue on the Geo-Environmental Problems Caused by Underground Construction

Bing Bai

School of Civil Engineering, Beijing Jiaotong University, Beijing 100044, China; bbai@bjtu.edu.cn

Geo-environmental disturbances refer to the significant changes in physical, mechanical, and even chemical properties of soils; are closely related to interrelated multi-physical field coupling systems of solid particles, water, and gas in the shallow stratum; and are caused by underground engineering construction. The evaluation of the possible geo-environmental hazards is an important, noteworthy topic that is relevant to researchers and engineers involved in the construction of large-scale underground engineering. This is related to the suitability of civil engineering construction, especially for some complicated geological conditions such as saturated super soft soil, laterite, loess, sand gravel stratum, saline soil, frozen soil, and karst stratum with rich water.

The goal of this Special Issue is to present the recent developments in the interaction between the coupled multi-physical fields and underground structures, as well as the geo-environmental effect, to stimulate fruitful technical and scientific exchange between professionals.

This Special Issue features nine papers covering the various fields of geo-environmental problems caused by underground construction. Zhao et al. [1] carried out some field monitoring and numerical simulation work in super-large span twin tunnels and discussed the vault settlement, the stress of concrete, and the sum pressure. Zhang et al. [2] used the GEO-Studio finite element software to explore the influence of the comprehensive slope rate on the permanent displacement when the slope rate of each grade of multi-stage loess slope changes and the stage of multi-stage slope changes. Additionally, Su et al. [3] completed a large model test of a foundation pit supported by a pile anchor with a geometric similarity ratio of 1:10. In their paper, the researchers discussed the force and deformation characteristics of the support structure by simulating the conditions of additional load at the pit edge, soil layered excavated, and anchors tensioned. Zhang et al. [4] derived calculation formulas of stability factors under the four arc slip surfaces of filled slopes reinforced by a frame with prestressed anchor plates by using the improved Bishop method and proposed a search method for the most dangerous slip surface. Additionally, Zhao et al. [5] conducted some capillary rise tests on soil columns containing three layers of sandy soils with coarser over finer over coarser sandy soil to investigate the effect of the relatively finer soil interlayer. Jiang et al. [6] reported the results from a field investigation to determine the influence of groundwater in the process of tunnel excavation and established a hydrogeological model of the region from the inverted regional natural flow field parameters. Meanwhile, Wang et al. [7] used a numerical simulation method to establish a model of multi-seam goafs with different spacing conditions to investigate the subsidence reduction effects of various grouting schemes. Baraibar et al. [8] discussed a special procedure for injecting cement and microcement to waterproof the surrounding drilling area, thus preventing tunnels from functioning as a drain for an aquifer. Lastly, Wang et al. [9] numerically simulated four semicircular stratified sandstone specimens with different strengths and seven different bedding angles using RFPA2D-Basic V2.0 software—these research results can provide a theoretical reference for the safety and stability of underground engineering. Overall, the research presented in this Special Issue will enrich the perspective on the disturbance and environmental effects of geotechnical engineering construction.

Citation: Bai, B. Special Issue on the Geo-Environmental Problems Caused by Underground Construction. *Appl. Sci.* **2023**, *13*, 10754. https://doi.org/10.3390/app131910754

Received: 22 September 2023
Accepted: 26 September 2023
Published: 27 September 2023

Although the submissions for this Special Issue have been closed, more in-depth research is still needed in the field of geo-environmental and construction disturbance to continue to address the many challenges that still need answers, such as construction disturbance effect, geo-environmental problems of special soils, environmental influence of energy pile, and seepage under multi-field coupling.

Funding: This research received no external funding.

Acknowledgments: Thanks to all the authors and peer reviewers for their valuable contributions to this Special Issue, "Geo-Environmental Problems Caused by Underground Construction". I would also like to express my gratitude to all the staff and people involved in this Special Issue.

Conflicts of Interest: The author declares no conflict of interest.

References

1. Zhao, X.; Sun, K.; Zhen, Y.; Hong, Y.; Zhou, H. Study on share rate of supports structure for super-large span twin tunnels with small interval. *Appl. Sci.* **2022**, *12*, 7498. [CrossRef]
2. Zhang, X.; Zhang, X.; Ye, S. Calculation for permanent displacement of single slip surface of multi-stage loess slope based on energy method. *Appl. Sci.* **2022**, *12*, 8426. [CrossRef]
3. Su, T.; Zhou, Y.; Wang, Z.; Ye, S. Large scale model test study of foundation pit supported by pile anchors. *Appl. Sci.* **2022**, *12*, 9792. [CrossRef]
4. Zhang, J.; Li, W.; Ye, S. Stability analysis of filled-slope reinforced by frame with prestressed anchor-plates under static action. *Appl. Sci.* **2023**, *13*, 1615. [CrossRef]
5. Zhao, Z.; Luo, Z.; Sun, H.; Li, H.; Liu, Q.; Liu, H. Capillary rise in layered soils. *Appl. Sci.* **2023**, *13*, 3374. [CrossRef]
6. Jiang, T.; Pei, X.; Wang, W.; Li, L.; Guo, S. Effects of the excavation of a hydraulic tunnel on groundwater at the Wuyue pumped storage power station. *Appl. Sci.* **2023**, *13*, 5196. [CrossRef]
7. Wang, H.; Qin, Y.; Guo, Y.; Xu, N. A Numerical simulation of the subsidence reduction effect of different grouting schemes in multi-coal seam goafs. *Appl. Sci.* **2023**, *13*, 5522. [CrossRef]
8. Baraibar, J.M.; Gil, M.; Escobal, I. Influence of the construction of the urdinbide road tunnel on the autzagane aquifer in Biscay (Spain). *Appl. Sci.* **2023**, *13*, 7034. [CrossRef]
9. Wang, G.; Meng, T.; Feng, G.; Ma, L.; Yang, W.; He, Y.; Zhang, Z.; Liang, X. Influence of bedding strength and angle on fracture characteristics of sandstone under three-point bending conditions. *Appl. Sci.* **2023**, *13*, 8216. [CrossRef]

applied sciences

Article

Deformation Law of Tunnels Using Double-Sidewall Guide Pit Method under Different Excavation Sequences

Yanmei Ruan [1], Xu Luo [1], Jin Li [2], Yang Li [3], Shan Lin [1], Chengkun Ling [2] and Bingxiang Yuan [3,*]

[1] Guangzhou Metro Design & Research Institute Co., Ltd., No. 204, Huan Shi Xi Road, Yuexiu District, Guangzhou 510010, China; ruanyanmei@gmdi.cn (Y.R.); luoxu@gmdi.cn (X.L.); linshan@gmdi.cn (S.L.)
[2] China Railway Tenth Bureau Group Urban Rail Transit Engineering Co., Ltd., No. 555, North Panyu Avenue, Donghuan Street, Panyu District, Guangzhou 511400, China; lijinxxx730@sohu.com (J.L.); lingchengkun@Icloud.com (C.L.)
[3] School of Civil and Transportation Engineering, Guangdong University of Technology, No. 100, Outer Ring West Road, Guangzhou University Town, Panyu District, Guangzhou 510006, China; 2112209148@mail2.gdut.edu.cn
* Correspondence: yuanbx@gdut.edu.cn

Abstract: The double-sidewall guide pit method finds extensive application in the construction of large cross-section tunnels in soft rock strata due to its minimal disruption to the surrounding rock, thereby enhancing tunnel stability. To investigate the loading and deformation patterns of the surrounding rock and tunnel support using the double-sidewall guide pit method, this study compares the impacts of various construction sequences on surface settlement, surrounding rock stress, and lining stress using indoor model tests. The experimental results show that after excavating the upper guide hole on one side, the excavation of the lower guide hole on the same side is carried out. The upper and lower support structures form a closed loop, and the structure can better constrain the surrounding rock and control the deformation of the surrounding rock, whereas the lower structure can share the stresses suffered by the upper structure. Therefore, compared with the upper and lower excavation methods, the surface settlement caused by the left and right excavation methods is smaller, the disturbance to the surrounding rock is smaller, and the supporting structure is more evenly and stably stressed in the excavation process.

Keywords: double-sidewall guide pit method; excavation sequence; modeling experiments

Citation: Ruan, Y.; Luo, X.; Li, J.; Li, Y.; Lin, S.; Ling, C.; Yuan, B. Deformation Law of Tunnels Using Double-Sidewall Guide Pit Method under Different Excavation Sequences. *Appl. Sci.* **2023**, *13*, 12764. https://doi.org/10.3390/app132312764

Academic Editor: Bing Bai

Received: 14 September 2023
Revised: 13 October 2023
Accepted: 2 November 2023
Published: 28 November 2023

1. Introduction

In recent years, the double-sidewall guide pit method has emerged as a crucial construction approach for mitigating disturbances to adjacent structures and tunnel deformations. This method addresses challenges posed by the dense urban population, extensive surface buildings, and the substantial infrastructure of subway pipe foundations and lines. Compared with other tunnel excavation methods, the double-sidewall guide pit method has great advantages in controlling surface settlement and horizontal displacement and is suitable for the construction of large cross-section tunnels [1–3]. Different excavation sequences of guide pits will have different effects on surface settlement, surrounding rock stresses, and lining stresses.

There are many factors affecting the construction of the double-sidewall-guide pit method. In terms of theory, Zhao et al. [4–6] studied the effect of the double-sidewall-guide pit method on surface settlement, vault settlement, and horizontal deformation and concluded that the tensile stress is dominant in the connection part of the initial support and diaphragm wall and the compressive stress is dominant in the middle and lower parts of the support structure. Liu et al. [7] studied the mechanical response of the support structure in the construction of a double-sidewall-guide pit in Class V enclosing rock. Liu et al. [8–10] conducted on-site monitoring and analysis of the large cross-section double-sidewall tunnels using the double-sidewall guide pit method and concluded that the deformation and force of the surrounding rock and lining are mainly directly related to the construction process and geological conditions used. Cui et al. [11,12] explored how the mechanical behavior of the surrounding rock palm surface changed with the construction sequence and determined and elucidated the influence of the construction sequence on the surface displacement caused by a tunnel excavation. Ling et al. [13–16] optimized the excavation sequence of the cross-section of tunnels using the double-sidewall guide pit method according to an actual situation on site, and the study showed that the support structure induced by the excavation sequence after the optimization was very good. Wang et al. [17,18] optimized the core column width of tunnels using the double-sidewall guide pit method, and the analysis showed that there was an optimal value for the core column width that could improve the stability of the surrounding rock. Jiao and Yang et al. [19–25] optimized construction parameters, compared the effects of different curvature radii and excavation schemes on the stability of surrounding rock, and selected the best construction scheme. Li et al. [26–28] optimized the construction sequence of the double-sidewall guide pit method, adapted the excavation form of the upper and lower sections, and preserved the core rock columns in the middle soil layer, which accelerated their construction progress and improved the overall stability of the tunnel.

This paper, building upon prior research findings, undertakes a comparative analysis of surface settlement, surrounding rock stress, and lining stress resulting from diverse construction sequences. The investigation employs indoor model tests to elucidate the mechanical response characteristics of both the surrounding rock and supporting structure during tunnel excavation under various construction sequences.

2. Introduction to the Experiment

2.1. Test Material and Apparatus Arrangement

This paper derives from a practical construction project and its test prototype parameters, as detailed in Table 1. To enhance the simulation of an authentic tunnel construction process and capture the effects of variations in the mechanical behavior of peripheral rock support, the indoor physical model test is conducted at a scale of 1:45. The similarity ratios for stress, elastic modulus, and cohesive force can be deduced by integrating the three fundamental theorems of similarity theory with the basic equations of elasticity: $C_\sigma = C_E = C_C = 45$. The displacement similarity ratio is $C_\delta = 45$. The strain similarity ratio is $C_\varepsilon = 1$. The internal friction angle similarity ratio is $C_\varphi = 1$.

Table 1. Physico-mechanical parameters of a moderately weathered rock stratum.

Stratigraphy	Physical Indicators			Mechanical Indicators	
Moderately weathered muddy siltstone	Natural Density (kN/m^3)	Modulus of elasticity/(MPa)	Poisson's ratio	Cohesion (kPa)	Angle of internal friction (°)
	23.5	1758	0.34	654	33.21

Following an examination of analogous materials in the surrounding rock, quartz sand was selected as the filling material, whereas lower-strength materials like clay or barite powder were chosen as the cementing material. Multiple sets of straight shear tests were conducted, and using iterative comparisons and analyses of model materials, the proportions of quartz sand, clay or barite powder, lubricant, and water were adjusted within a defined range to establish mechanical indices. The basic physical parameters of the tested soil are presented in Table 2.

Table 2. Experimental soil ratio and physical parameters.

Quartz Sand	Clays	Lubricants	Water	Cohesion (kPa)	Angle of Internal Friction (°)
47%	8%	1%	14%	12.70	32.33

Following multiple tests, it was observed that the lining composed of barite powder exhibited a low modulus of elasticity and failed to meet the required similarity ratio. Consequently, the lining model material was altered to a mixture of gypsum, water, and a small amount of cement. The modulus of elasticity of the lining's analogous materials varies with the mass ratio of gypsum, water, and cement. To achieve the stipulated similarity ratio, unconfined uniaxial compressive strength tests were conducted on the lining's similar materials with different mix ratios, utilizing an electronic universal testing machine. The stress–strain curves and modulus of elasticity were measured, and after several tests, it was determined that the gypsum: water: cement ratio of 1:0.75:0.25 met the similarity requirements. The finalized mix yielded a similar material modulus of elasticity of 558.66 MPa.

The primary objective of this test is to analyze the loading and deformation patterns of perimeter rock support induced by various construction sequences. The measured parameters encompass surface deformation, perimeter rock stress, and lining stress. Surface deformation is quantified using a percentage meter with a range of 0–10 mm and an accuracy of 0.01 mm, as depicted in Figure 1. The DMTY-type resistive strain earth pressure box, illustrated in Figure 2, is employed to gauge static or dynamic earth pressure values in the surrounding rock. Data on the stresses of both the surrounding rock and the lining are collected using the DH3818Y static strain tester, as shown in Figure 3.

Figure 1. Dial indicator.

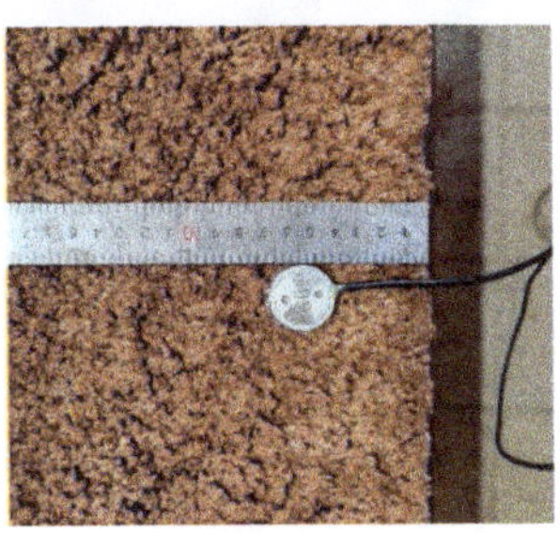

Figure 2. Embedded soil pressure cell.

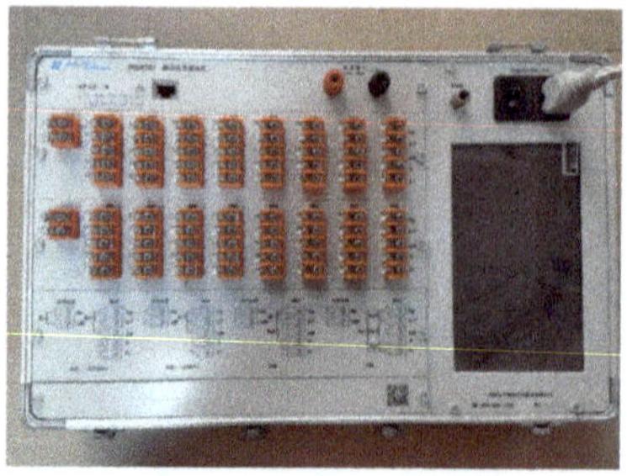

Figure 3. DH3818Y static strain tester.

2.2. Model Experimental Setup

Indoor model tests are used to study field engineering problems with a certain scaling based on the similarity theory. Geotechnical engineering, especially underground structural engineering, has complex geological conditions and many uncertainties in engineering, so it is still difficult to accurately solve the structural forces in mathematical calculation and mechanical analysis. Indoor model tests can remove mathematical and mechanical difficulties using real physical entities, and reflect the physical and dynamic response of the surrounding rock support structure in the process of tunnel excavation in a detailed, real, and direct way [29–31].

The geometric scale for this indoor physical modeling test is 1:45, and the vertical loading is achieved using a custom-made steel frame and model box. The overall dimensions of the model box are 180 × 150 × 50 cm (length × height × thickness), as illustrated in Figure 4. The frame of the model box consists of impermeable square steel panels with a thickness of 30 mm, which are spliced together. Positioned in front of the frame is a 10 mm thick acrylic glass plate, while the remaining sides are constructed with 10 mm thick wooden boards, covered with a layer of rubber for effective sealing. In accordance with the specifications of the actual project, the upper part of the tunnel features 65 cm thick surrounding rock, and the lower part has 50 cm thick surrounding rock.

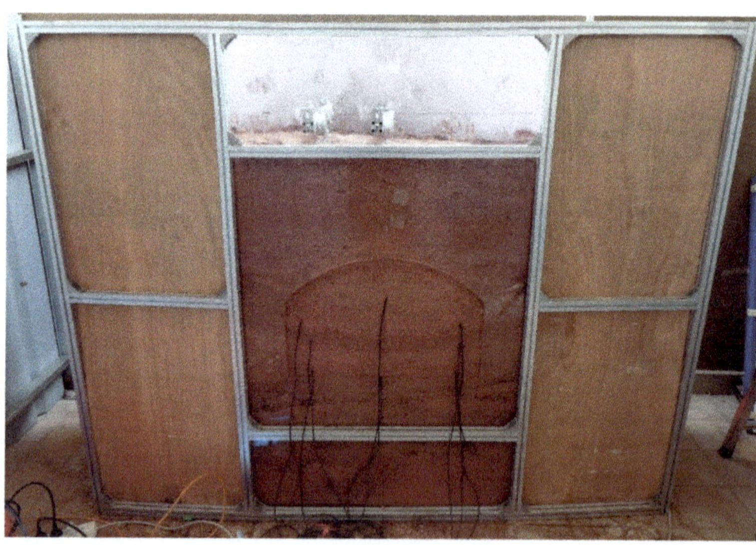

Figure 4. Overall model box.

2.3. Experimental Program and Experimental Macrophenomena

This test replicates the excavation process of a tunnel using the double sidewall guide pit method, dividing the tunnel into six guide pits, as illustrated in Figure 5. The experiment is categorized into two groups: the first group, an excavation sequence optimization group, employs the up-and-down excavation method. It is initiated by excavating the upper left and right guide holes, followed by the upper intermediate guide holes. Subsequently, the lower left and right guide holes are excavated, concluding with the excavation of the lower intermediate guide holes, sequenced as ①–②–③–④–⑤–⑥. The second group follows the left-and-right excavation method, where the left and right cross sections are initially constructed to promptly form the support system, and the intermediate section is subsequently developed, sequenced as ①–④–②–⑤–③–⑥.

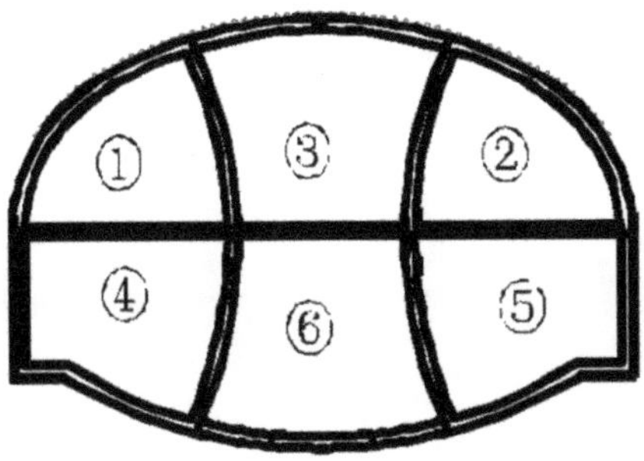

Figure 5. Double-side drift method and tunnel guide hole number.

Following the determination of the matching ratio from the proportioning test, the perimeter rock similar material is meticulously prepared. The soil body is then filled to the specified height, with the preburial of the tunnel lining model. Subsequently, the surrounding soil is compacted, ensuring full contact between the outer surface of the lining and the soil body. Simultaneously, the earth pressure box is preburied at the designated position, and the perimeter rock similar materials are left undisturbed for 12 h after filling to ensure that the perimeter rock deformation, soil solidification, and settlement essentially stabilize. After achieving stabilization, the experimental setup proceeds with the installation of a percentage meter to measure surface settlement. The initial reading is recorded after meter setup, and subsequent readings are noted before and after each step of the guide hole excavation. A strain signal acquisition and analysis system is established in the computer, capturing the initial values of the earth pressure box and the strain gauge. Simultaneously, the software is configured to automatically save data, facilitating continuous data collection from the earth pressure box and the strain gauge. Guide holes are then excavated in different sequences, and the temporary support of each section is sequentially dismantled after all guide holes have been excavated. An interval of 12 h is maintained between each support dismantling, followed by an additional 12-h period of standing time after the removal of all temporary supports. After the stabilization of the soil body deformation, relevant test data are recorded. The entire experimental process is depicted in Figures 6–8.

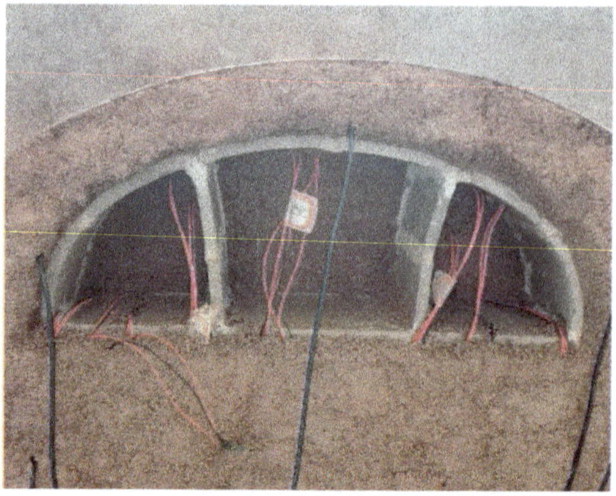

Figure 6. Excavating the upper guide hole.

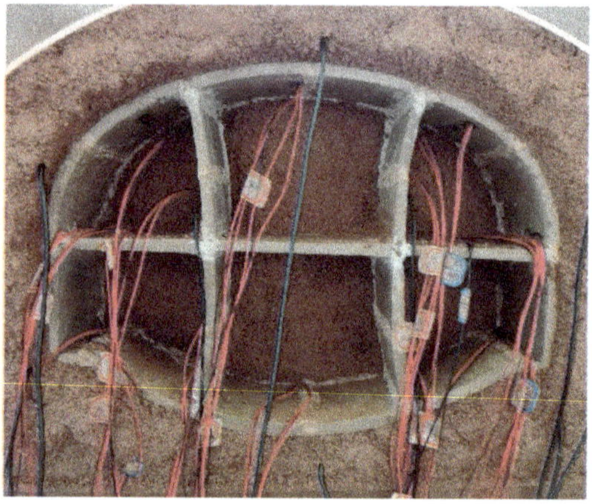

Figure 7. Guide hole excavation completed.

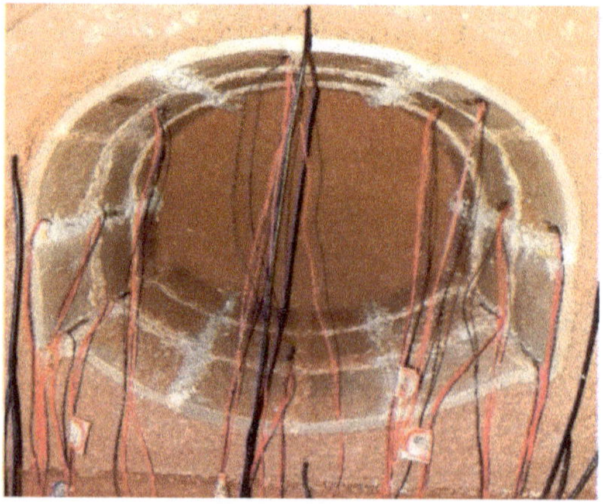

Figure 8. Completion of removing temporary tunnel supports.

3. Analysis of Test Data

3.1. Analysis of Surface Deformation Results

Figure 9 shows that the surface settlement above the center line of the tunnel is more pronounced compared to the left line. For lateral monitoring points, the final settlement value at monitoring point 1 is 0.238 mm, which is 10.70% larger than the final settlement value of 0.215 mm at monitoring point 4. The settlement at monitoring point 2 is 12.03% larger than that at monitoring point 5, and the settlement at monitoring point 3 is 26.09% larger than that at monitoring point 6. This discrepancy arises from the multiple construction disturbances of the left and right guide holes and the middle guide hole affecting the surface above the center line, resulting in greater settlement. In contrast, the left line experiences a relatively smaller impact from the excavation of the left and middle guide holes, with the excavation of the right guide hole having less influence. Examining longitudinal monitoring points, the settlement at monitoring point 1 surpasses that at monitoring point 2 by 34.46%, monitoring point 2 exceeds monitoring point 3 by 205.17%, monitoring point 4 surpasses monitoring point 5 by 36.06%, and monitoring point 5 exceeds monitoring point 6 by 243.48%. This pattern indicates that the closer the location to the opening, the larger the surface settlement, while locations further from the opening exhibit smaller settlement values. The disturbance caused by the excavation of the first guide hole affects the surface of the cave entrance, and subsequent guide hole excavations continue to impact the surface, resulting in larger settlement values closer to the entrance. In this test, monitoring points 5 and 6, located at the surface of the unexcavated soil behind the tunnel, experienced minimal settlement.

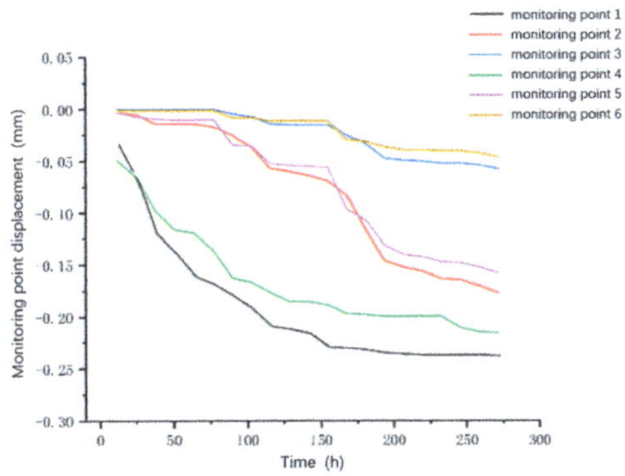

Figure 9. Surface deformation time history curve of the first group.

While the excavation of the upper guideway exerts a more pronounced impact on surface settlement, the timely application of the upper guideway to the support structure following the completion of excavation serves to support and restrict the vertical displacement of the soil body. Consequently, the excavation of the lower guideway has a comparatively smaller impact on surface settlement. Focusing on monitoring point 1 for analysis, the surface settlement pattern of the tunnel reveals that the excavation of the upper guide hole directly beneath monitoring point 1 exerts the most significant influence on its settlement. Monitoring point 1 settles by 0.034 mm, 0.034 mm, and 0.052 mm after the excavation of the upper-left 1, upper-right 1, and upper-right 1 guide holes, respectively, accounting for 14.29%, 14.29%, and 21.85% of the total settlement value. The excavation of the upper guideway represents a crucial construction phase necessitating prompt application of support structures and vigilant monitoring. Subsequent excavation of the lower guideway has a diminished impact due to the support and restraint provided by the upper superstructure. The settlement from the excavation of the second and third ring sections amounts to 0.060 mm and 0.008 mm, respectively, constituting 25.21% and 3.36% of the total settlement value.

The general trend of surface settlement at monitoring point 1 indicates that in the early stage when the excavation surface is directly below the monitoring point, settlement drops rapidly with the largest settlement rate. As the excavation surface advances and gradually moves away from the monitoring point, the settlement rate diminishes, and the settlement curve gradually flattens out, ultimately stabilizing. Specific surface deformation data for the second group of tests are depicted in Figure 10. The surface settlement trend caused by different construction sequences exhibits slight variations, yet the overall pattern remains consistent. Settlement is primarily influenced by the excavation of the upper guide hole and the distance between the excavation surface. The second group of the test excavation program involves constructing the left and right side sections first and subsequently excavating the middle section after the upper and lower support structures form a closed loop. Settlement values at monitoring points 1–6 are, respectively, 4.62%, 18.64%, −1.70%, 13.49%, 6.96%, and 58.70% smaller than those of the first group of tests, indicating that the timely formation of the closed loop of the support structure has a certain control effect on surface settlement.

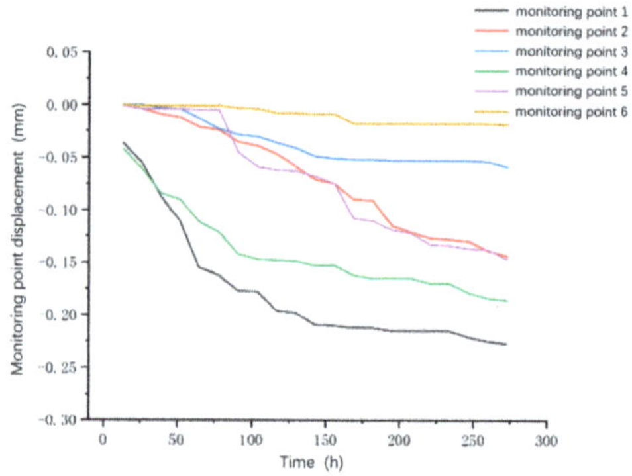

Figure 10. Surface deformation time history curve of the second group.

3.2. Analysis of Surrounding Rock Stress Results

The stress-time curve of the surrounding rock in the first test group is depicted in Figure 11. The examination of the figure reveals that the peripheral rock at each point around the tunnel, under the initial stress state, is subjected to compressive stress. The magnitude of the stress is directly correlated with the depth of burial, with greater depths resulting in higher initial stress values. The initial stress value of the peripheral rock at the arch top is 7.449 kPa. The average initial stress value of the peripheral rock at the arch shoulder is 8.497 kPa. The average initial stress value of the peripheral rock at the arch waist is 8.646 kPa. The average initial stress value of the peripheral rock at the foot of the arch is 9.079 kPa. and the average initial stress value of the peripheral rock at the bottom of the arch is 10.113 kPa.

After the excavation of the upper middle 1 guide hole, the peripheral rock at the arch top undergoes a rapid release of the original stress, resulting in a swift decrease in compressive stress from 7.188 kPa to 2.870 kPa. Subsequently, the stress stabilizes under the influence of the supporting structure. The upper guide hole of the second ring, situated adjacent to the peripheral rock of the arch, experiences a gradual increase in peripheral rock stress, as the stress released after excavation is transferred to the arch. The stress value after the completion of the excavation of the middle and upper 2 guide holes reaches 3.680 kPa. The excavation of the lower guide hole of the second ring and the third ring section exerts a lesser influence on peripheral rock stress, leading to a slow and fluctuating increase in stress. The stress value after the completion of the excavation of all guide holes amounts to 4.213 kPa. Upon the removal of the temporary support, the peripheral rock stress increases due to disturbance, resulting in a compressive stress value of 5.241 kPa when all temporary supports are removed.

The stress distribution of the left and right surrounding rock exhibits symmetry, and the stress variation pattern is essentially identical. Consequently, the arch shoulder, arch waist, and arch foot are collectively considered as the left surrounding rock for analysis. Following the excavation of the upper left 1 guide hole, the compressive stress undergoes a rapid decline from 8.342 kPa to 4.860 kPa. Subsequent excavation of the upper left 2 guide holes leads to a slight increase in stress to 5.701 kPa, after which the stress value stabilizes. Minimal fluctuations occur in the subsequent excavation phases, and upon the completion of all guide holes, the stress value reaches 6.035 kPa. The removal of the temporary support induces fluctuations in the surrounding rock stress value due to disturbances. Specifically, when all temporary supports are removed, the compressive stress of the surrounding rock fluctuates, settling at a value of 5.916 kPa.

The arch-waist perimeter rock experiences stress release after the excavation of both upper-left 1 and lower-left 1 guide holes, resulting in a decrease in stress values. Specifically, during the excavation of the upper-left 1 guide hole, the stress decreases from 8.854 kPa to 7.600 kPa, and during the excavation of the lower-left 1 guide hole, it decreases from 8.175 kPa to 7.514 kPa. Subsequent excavation of the adjacent upper-left 2 and lower-left 2 guide holes induces a gradual increase in stress values, reaching from 7.401 kPa to 7.748 kPa during the excavation of upper-left 2 guide hole and from 7.810 kPa to 8.082 kPa during the excavation of lower-left 2 guide hole. Upon the removal of the temporary support, the initial support undergoes inward shrinkage and deformation due to the lack of support. At the arch waist, the surrounding rock briefly forms a critical surface, causing a decrease in stress values. The compressive stress of the surrounding rock settles at a value of 6.431 kPa upon the completion of the removal of temporary support.

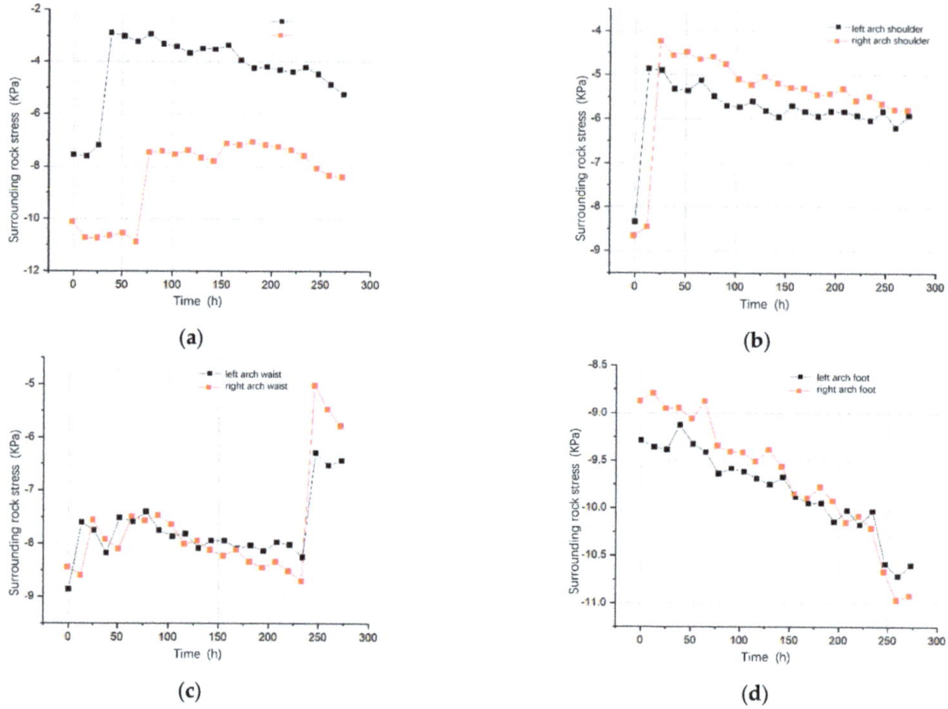

Figure 11. The time-history curve of surrounding rock stress in the first group of the test. (**a**) Stress time curve of the surrounding rock at the top and bottom of the arch. (**b**) Stress time curve of the surrounding rock at the arch shoulder. (**c**) Stress time-course curve of the perimeter rock at the arch girdle. (**d**) Stress time curve of the surrounding rock at the foot of the arch.

The peripheral rock at the foot of the arch, situated below the lower left guide hole, exhibits an overall gradual increase in stress values. The peripheral rock stress value reaches 10.033 kPa upon the completion of all guide hole excavations and further increases to 10.601 kPa after the removal of all temporary supports.

Additionally, the peripheral rock at the bottom of the arch undergoes a stress release after the excavation of the middle-lower 1 guide hole and middle-lower 2 guide holes. The stress value reduces to 7.458 kPa following the excavation of the middle-lower 1 guide hole and further decreases to 7.098 kPa after the excavation of the middle-lower 2 guide hole. Subsequent excavation of the middle and lower 2 guide holes leads to a decrease in stress values from 10.886 kPa to 7.458 kPa and from 7.786 kPa to 7.098 kPa, respectively. The remaining guide hole excavations result in a gradual increase in stress values. Upon the completion of all guide hole excavations, the stress value of the surrounding rock settles at 7.572 kPa. After removing all temporary supports, the stress value of the entire structure reaches 8.381 kPa.

The observed stress change patterns in the surrounding rock reveal that, following guide hole excavation, the area near the excavation zone experiences deformation toward the excavated chamber due to the release of constraints. During this deformation process, a portion of the energy is released, leading to a reduction in compressive stress. The diminished energy is then transferred to the surrounding geotechnical body, resulting in the redistribution of stress within the surrounding rock. Upon the application of the support structure, it generates resistance against the movement of the rock body, establishing corresponding constraints. Consequently, the deformation of the surrounding rock gradually weakens until a new balance is achieved between the surrounding rock and the support structure. Upon the removal of the diaphragm wall providing vertical support and the temporarily elevated arch providing horizontal support, the joint between the initial support and the temporary support undergoes inward shrinkage and deformation due to the lack of support. In this scenario, the nearby surrounding rock temporarily forms a critical surface, leading to a reduction in stress values that are then transferred to the surrounding area.

The specific data on surrounding rock stress in the second group of tests are presented in Figure 12. It is evident from the figure that, due to the different excavation sequences in the second group of tests, the temporal changes in perimeter rock stress at each point differ from those in the first group. Nonetheless, the overall pattern remains consistent. The perimeter rock stress is primarily influenced by the excavation of adjacent guide holes, causing a decrease in stress values near the excavation area and an increase in neighboring perimeter rock stress values. The construction of guide holes at a greater distance has a lesser impact on perimeter rock stress values. To gauge the degree of disturbance of excavation on the surrounding rock, average stress change values were calculated. A larger value indicates a greater disturbance caused by excavation. The average stress change values at the top of the arch, the left and right arch shoulders, the left and right arch waist, the left arch foot, and the bottom of the arch are smaller in the second group by 2.81%, −9.66%, 12.25%, 23.59%, 38.35%, 22.04%, and 16.76%, respectively, compared to the first group. This suggests that the left and right excavation methods result in less disturbance to the surrounding rock. This is attributed to the excavation of upper and lower guide holes on the same side first, forming a closed loop in the upper and lower supporting structures, enabling better restraint of the surrounding rock. Consequently, the deformation of the surrounding rock during excavation is minimized, resulting in smaller changes in stress values.

3.3. Analysis of Lining Stress Results

Figure 13 illustrates the time course curve of lining stress for the first set of tests. Following the excavation of the upper left 1 guide hole, the surrounding rock undergoes deformation towards the excavation chamber. The supporting structure, in turn, bears the force exerted by the surrounding rock. Consequently, the lining stress values of the left arch shoulder, left arch waist, left elevation arch, and the upper left diaphragm wall all experience an increase. Specifically, the stress value of the left arch waist rises from 2.454 kPa to 7.174 kPa, the left elevation arch stress increases from 0.158 kPa to 1.456 kPa, and the stress value of the upper left diaphragm wall rises from −0.973 kPa to 7.730 kPa.

After excavating the upper-right 1 guide hole, the lining stress values for the right arch shoulder, right arch waist, right elevated arch, and upper-right diaphragm wall all experience an increase. Specifically, the stress value of the right arch waist rises from −0.251 kPa to 5.328 kPa, the stress value of the right elevated arch increases from 0.884 kPa to 1.261 kPa, and that of the upper-right diaphragm wall rises from 0.579 kPa to 8.096 kPa. The left side of the supporting structure is relatively unaffected by the excavation of the upper-right 1 guide hole, with stress values showing only slight fluctuations. Following the excavation of the middle-upper 1 guide hole, the overall superstructure experiences an elevation in stress values. The lining stress of the arch top increases from 1.103 kPa to 5.556 kPa, the left arch waist stress rises to 9.583 kPa, the right arch waist stress increases to 7.239 kPa, and the left superelevation arch stress reaches 2.679 kPa. The stress values of the right superelevation arch and left upper diaphragm wall rise to 1.675 kPa and 9.908 kPa, respectively, while the right upper diaphragm wall stress increases to 11.432 kPa. This indicates that the applied vault lining and the middle superelevation arch connect the two sides of the supporting structure into a unified whole, transferring the force exerted by the overlying surrounding rock to the entire superstructure. Consequently, the overall stress value of the structure increases. The arch top and diaphragm wall bear the direct vertical soil pressure, resulting in higher stress values compared to the rest of the structure. The stress values of the right arch waist and right upper diaphragm wall decrease to 2.427 kPa and 10.936 kPa, respectively, while the stress values of the right foot of the arch, the right lower diaphragm wall, and the right superelevation arch rise to 1.126 kPa, 6.924 kPa, and 3.377 kPa, respectively. Construction disturbances lead to damage in the support structure of the remaining guideway, causing stress values to fluctuate. Following the excavation of the middle and lower 1 guide hole, stress values for the upper left diaphragm wall, upper right diaphragm wall, left elevation arch, and right elevation arch decrease to 7.529 kPa, 8.142 kPa, 2.761 kPa, and 2.486 kPa, respectively. Simultaneously, stress values for the left arch waist, right arch waist, left arch foot, right arch foot, lower left diaphragm wall, lower right diaphragm wall, and the bottom of the arch rise to 5.524 kPa, 4.374 kPa, 2.178 kPa, 2.484 kPa, 10.485 kPa, 10.468 kPa, and 3.179 kPa. This indicates that after the excavation of the middle and lower guiding tunnel and the application of the arch bottom supporting structure, the supporting structure of the first ring tunnel connects into a cohesive whole, leading to redistributed and downward concentrated lining stresses.

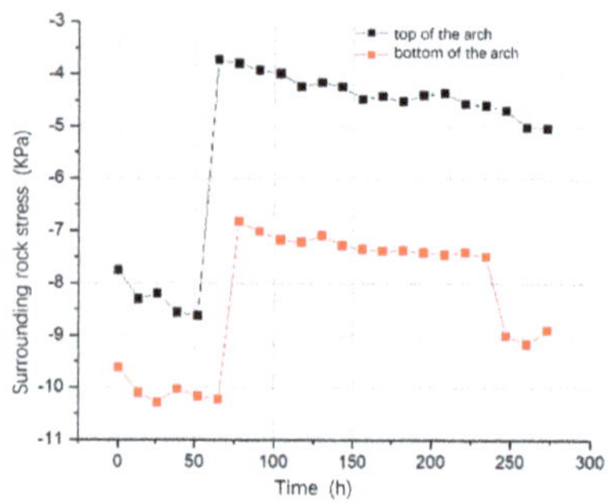

(**a**)

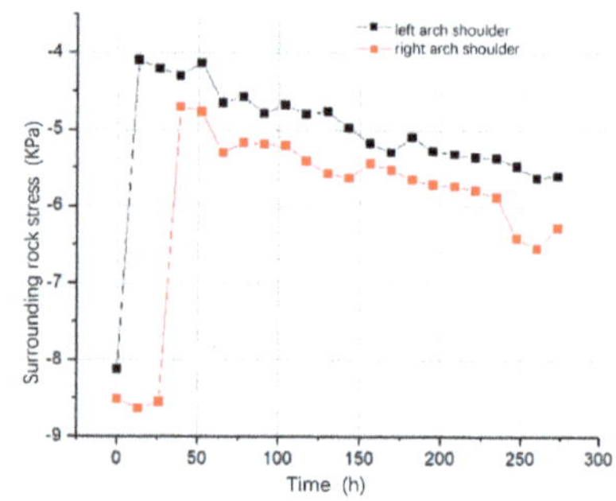

(**b**)

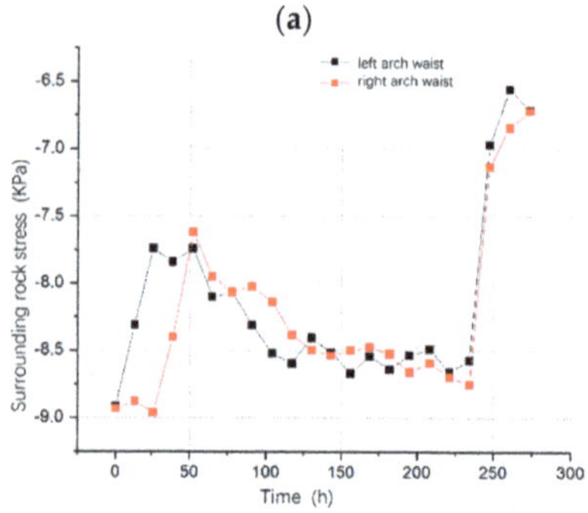

(**c**)

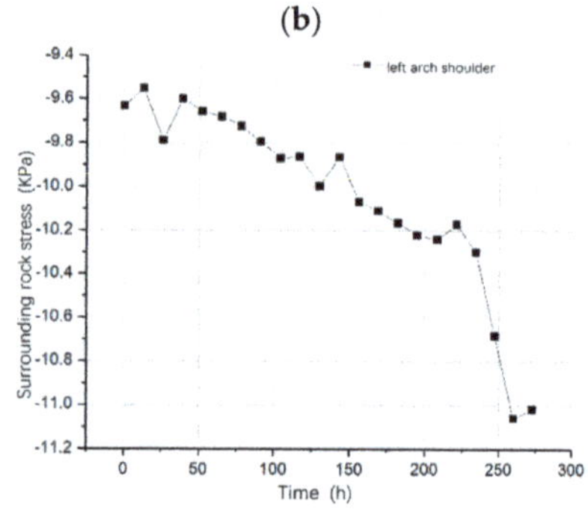

(**d**)

Figure 12. The time-history curve of surrounding rock stress in the second group of the test. (**a**) Stress time curve of the surrounding rock at the top and bottom of the arch. (**b**) Stress time curve of the surrounding rock at the arch shoulder. (**c**) Stress time-course curve of the perimeter rock at the arch girdle. (**d**) Stress time curve of the surrounding rock at the foot of the arch.

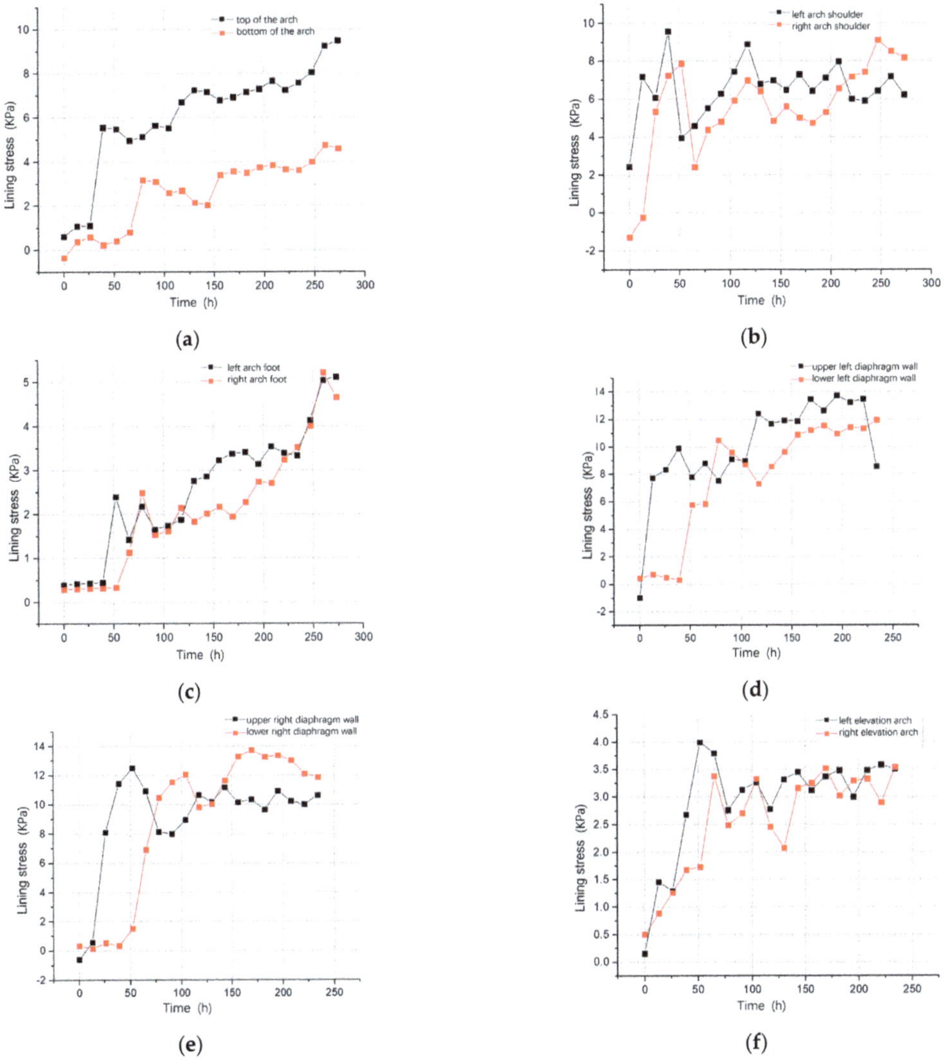

Figure 13. The time-history curve of lining stress in the first group of the test. (**a**) Stress time curve of the lining of the vault and the base of the arch. (**b**) Stress time curve of arch waist lining. (**c**) Stress time-course curve of footing lining. (**d**) Stress time curve of the left diaphragm wall lining. (**e**) Stress time-curve of the right-central diaphragm wall lining. (**f**) Temporary back arch lining stress time curve.

The construction of the entire upper guide tunnel for the second ring tunnel leads to increased stress values at various points. Specifically, the arch top, left arch waist, right arch waist, left upper diaphragm wall, and right upper diaphragm wall experience increases to 6.690 kPa, 8.894 kPa, 6.973 kPa, 12.405 kPa, and 10.663 kPa, respectively. Concurrently, the stress values of the left lower diaphragm wall and right lower diaphragm wall decrease to 7.287 kPa and 9.834 kPa. This shift suggests that stresses on the second ring support structure are transferred to the first ring, resulting in heightened stress on the upper structure of the first ring. Simultaneously, the lower structure of the second ring, which has not been applied yet, exhibits non-uniform stress distribution and force transmission, leading to upward stress concentration and a reduction in stresses on the lower support structure. Following the excavation of the second ring's lower guide hole, lining stresses re-center to the lower support structure. Stress values for the arch waist, upper left center diaphragm wall, and upper right center diaphragm wall decrease, while those for the arch foot, arch bottom, lower left center diaphragm wall, lower right center diaphragm wall, and superelevation arch increase.

The stress pattern of the support structure in the third ring section excavation mirrors that of the second ring, albeit with a smaller overall impact due to the increased distance from the first ring. The stress values of the first ring's support structure fluctuate up and down. Upon the completion of all guide hole excavations, the lower left and lower right diaphragm walls bear the highest stresses at 11.932 kPa and 11.841 kPa, respectively. They are followed by the upper left and right diaphragm walls at 8.536 kPa and 10.626 kPa, and then the arch roof at 7.568 kPa. The stress values for the arch top follow at 7.568 kPa, with the left and right arch waist stresses at 5.897 kPa and 7.415 kPa, respectively. Finally, the stress values for the left and right foot of the arch, the left and right superelevation arches, and the bottom of the arch show minimal differences, maintaining values near 3.503 kPa. This analysis reveals that the primary stress-bearing positions in the double sidewall tunnel are the diaphragm wall, arch top, and lining at the arch waist. The diaphragm wall handles the vertical load imposed by the overlying surrounding rock, while the vault and arch waist bear and transfer the vertical and horizontal loads from the surrounding rock. Monitoring the arch footing, positioned horizontally at the corner of the side wall, is crucial, as per the stress transfer principles mentioned above. Stress concentration is expected in the vertical arch footing of the side wall, making it a significant stress point in the supporting structure. Conversely, stresses in the horizontal arches, foot arch bottom, and temporary elevation arches are relatively small.

Upon the removal of temporary support, the stresses initially borne by the diaphragm wall and temporary elevated arch are transferred to the initial support, resulting in an increase in the stress values of the initial support. Following the removal of the temporary support from the first ring, the stresses on the diaphragm wall and temporary support shift to be supported by the initial support of the first ring and the temporary support of the second ring. The temporary support of the second ring contributes to the support capacity, causing a minor increase in the stress value of the initial support of the first ring. Specifically, the stress values of the arch top, left and right girdle, left and right foot of the arch, and arch bottom increase by 0.474 kPa, 0.524 kPa, 1.690 kPa, 0.803 kPa, 0.484 kPa, and 0.374 kPa, respectively. Upon the removal of the temporary support from the second ring, the initial support of the first ring assumes the majority of the pressure from the overlying peripheral rock, resulting in a significantly higher stress value. The stress values of the arch top, left and right arch waist, left and right footings, and the bottom of the arch increase by 1.210 kPa, 0.755 kPa, −0.590 kPa, 0.910 kPa, 1.210 kPa, and 0.751 kPa, respectively. The third ring's temporary support, which is distant from the initial support of the first ring, experiences minimal impact upon the removal of the first ring's initial support. The stress values show slight fluctuations, and the final stress values for the arch, left and right arch waist, left and right arch foot, and the bottom of the arch are 9.498 kPa, 6.221 kPa, 8.152 kPa, 5.120 kPa, 4.658 kPa, and 4.585 kPa, respectively.

Figure 14 illustrates the time course curve of lining stress for the second set of tests. Following the excavation of the upper left 1 guide hole, the stress values for the left arch waist, left elevated arch, and upper left diaphragm wall experience an increase, reaching 9.073 kPa, 0.817 kPa, and 8.703 kPa, respectively. Notably, the stress change is most significant for the left arch waist and the upper left diaphragm wall in this construction step. Consequently, it is crucial to enhance monitoring efforts for the left arch waist and the upper left diaphragm wall during the excavation of the upper left 1 guide hole. After excavating the lower left 1 guide hole, the lower structure assumes a portion of the stress, resulting in a decrease in stress values for the upper structure and an increase in stress values for the lower structure. Stress values for the left arch waist and the left upper diaphragm wall decrease to 3.696 kPa and 5.353 kPa, respectively. Simultaneously, stress values for the left arch foot, left elevated arch, and left lower diaphragm wall increase to 2.812 kPa, 5.990 kPa, and 3.248 kPa, respectively.

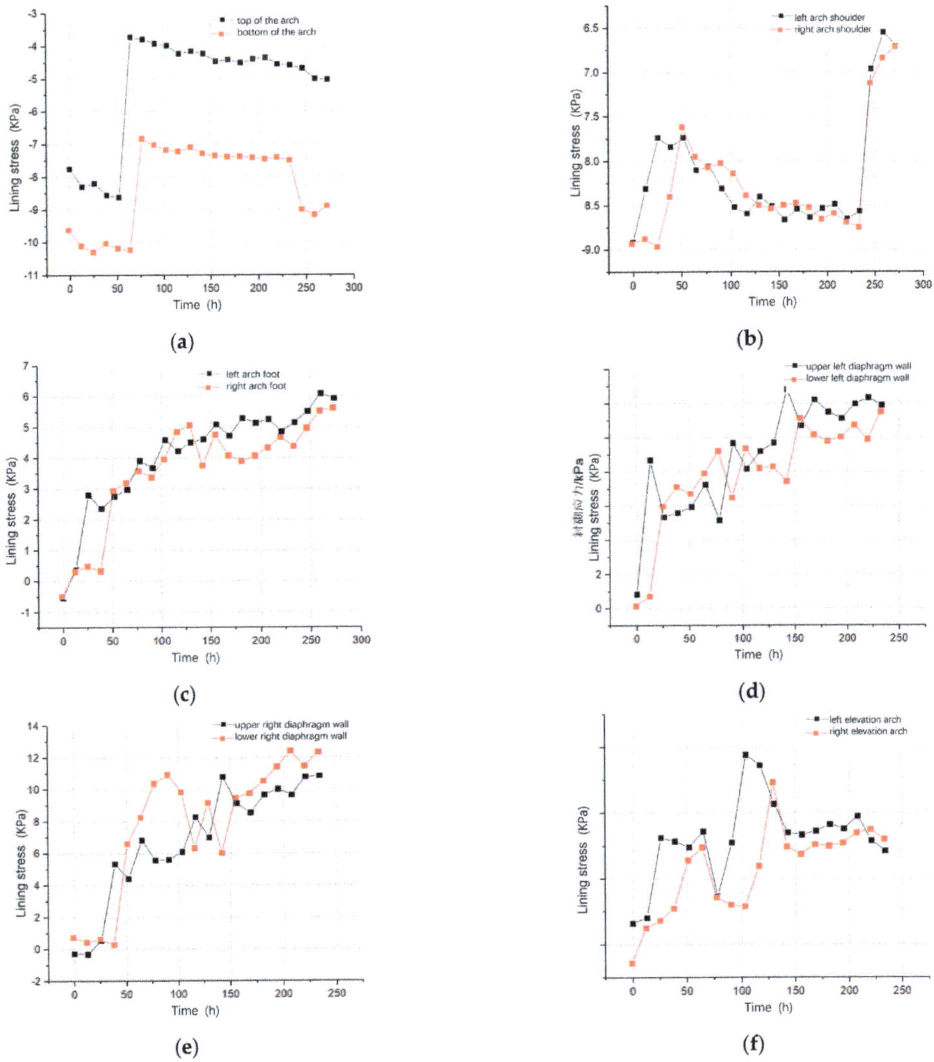

Figure 14. The time-history curve of lining stress in the second group of tests. (**a**) Stress time curve of the lining of the vault and the base of the arch. (**b**) Stress time curve of arch waist lining. (**c**) Stress time-course curve of footing lining. (**d**) Stress time curve of the left diaphragm wall lining. (**e**) Stress time-curve of the right-central diaphragm wall lining. (**f**) Temporary back arch lining stress time curve.

Excavating the upper right 1 guide hole induces an increase in stress values for the right arch waist, right elevation arch, and upper right diaphragm wall, reaching 6.813 kPa, 1.112 kPa, and 5.345 kPa, respectively, with minimal impact on the structure on the left side. Following the excavation of the lower right 1 guide hole, stress values for the right arch waist and the upper right diaphragm wall experience a rapid decrease to 2.952 kPa and

4.433 kPa, while stress values for the right arch foot, right elevation arch, and lower right diaphragm wall increase to 2.957 kPa, 2.576 kPa, and 6.618 kPa, respectively. The stress values on the left side structure show small fluctuations. Post-excavation of the middle and upper 1 guide holes, stresses on both sides of the supporting structure increase. However, the overall stress rise is not significant due to the applied substructure. The left and right arch foot, left and right arch waist, left upper diaphragm wall, left lower diaphragm wall, right upper diaphragm wall, right lower diaphragm wall, and left and right elevation arches experience elevation, reaching 4.526 kPa, 3.821 kPa, 2.992 kPa, 3.218 kPa, 7.267 kPa, 7.931 kPa, 6.836 kPa, 8.262 kPa, 3.437 kPa, and 2.965 kPa, respectively. Upon the excavation of the middle and lower 1 guideway, stress within the tunnel concentrates downward, resulting in a decrease in superstructure stress and an increase in substructure stress. The stress values for the upper left diaphragm wall, upper right diaphragm wall, left elevation arch, and right elevation arch decrease to 5.169 kPa, 5.576 kPa, 1.471 kPa, and 1.451 kPa, respectively. Meanwhile, the stress values for the left arch waist, right arch waist, left arch foot, right arch foot, lower left diaphragm wall, lower right diaphragm wall, and arch bottom increase to 5.606 kPa, 5.001 kPa, 3.915 kPa, 3.605 kPa, 9.244 kPa, 10.386 kPa, and 4.683 kPa.

The excavation of the left section of the second ring tunnel predominantly affects the support structure on the left side. The stress values for the left arch waist, left arch foot, left upper diaphragm wall, left lower diaphragm wall, and left superelevation arch increase to 5.954 kPa, 4.596 kPa, 8.194 kPa, 9.380 kPa, and 5.766 kPa, respectively, while the stress value on the right side of the support structure undergoes minimal change. Subsequently, the excavation of the right section also elevates the stress value of the right supporting structure. The stress values for the right arch waist, right arch foot, right upper diaphragm wall, right lower diaphragm wall, and right superelevation arch increase to 5.047 kPa, 5.078 kPa, 6.988 kPa, 9.170 kPa, and 4.932 kPa, respectively. The stress is concentrated again on the substructure after the completion of the intermediate section excavation.

The stress pattern of the support structure during the excavation of the third ring section follows the same trend as the second ring, albeit with a lesser degree of influence. Upon completing all guide hole excavations, the stress magnitude for each structure matches that of the first group. The highest to lowest stress values are the lower diaphragm wall, upper diaphragm wall, arch waist, arch bottom, arch foot, and superelevation arch, respectively. The stress change after removing the temporary support follows the same rule as the first group, resulting in an elevated stress value for the initial support. The final stress values for the left and right arch waist, left and right arch foot, and arch base are 7.741 kPa, 8.812 kPa, 5.934 kPa, 5.233 kPa, and 7.464 kPa, respectively.

Compared with the first set of tests, the second set was conducted after excavating the upper guide tunnel on one side, allowing subsequent excavation of the lower guide tunnel on the same side. This enabled the substructure to share some of the stresses, resulting in a more evenly distributed overall structural stress. In the first group of tests, the average stress values for the left arch waist and left upper diaphragm wall during the excavation of the first ring tunnel were 5.621 kPa and 7.018 kPa, respectively. In the second group of tests, these values decreased to 4.487 kPa and 5.621 kPa, respectively, representing a 20.17% and 19.91% reduction compared to the first group. Excavating the right (left) side guide tunnel followed by the excavation of the lower guide tunnel in both groups resulted in the connection of the upper and lower structures, forming a closed loop and increasing overall stability. Construction disturbance had minimal impact on the entire structure. In the first group, after excavating the upper-left 1 guide hole and applying the supporting structure, the construction of the upper-right 1 guide hole reduced the stress value of the left arch waist from 7.174 kPa to 6.073 kPa, with a strain change of 1.101 kPa. In the second group, after the construction of the lower-left 1 guide hole, the construction of the upper-right 1 guide hole and the lower-right 1 guide hole reduced the stress value of the left arch waist from 3.696 kPa to 3.550 kPa, and then increased to 3.914 kPa, with an average stress change of 0.255 kPa. Following the construction of both side guide holes and then the middle guide hole, the stress increase value for both side structures was smaller in the second group after the substructure was applied, allowing it to share the stress of the upper structure. In the first group, excavating the middle upper 1 guide hole increased the stress value of the left arch waist from 6.073 kPa to 9.583 kPa, with a stress change of 3.510 kPa. In the second group, excavating the middle upper 1 guide hole increased the stress value of the left arch waist from 3.914 kPa to 4.526 kPa, with a stress change of 0.621 kPa.

In summary, the left-and-right excavation method and the up-and-down excavation method result in little difference in the final support stress magnitude. However, using the left-right excavation method yields a more evenly distributed and stable support structure during the excavation process, minimizing the impact of construction disturbance.

4. Conclusions

In this paper, based on modeling tests, two sets of modeling tests were carried out using two different excavation sequences, the up-and-down excavation method and the left-and-right excavation method, to determine the interaction between the surrounding rock and the supporting structure by studying the surface settlement, the surrounding rock stresses, and the lining stresses induced by the different excavation sequences:

(1) For the surface settlement law, at the same burial depth, the surface settlement is mainly related to the excavation surface approaching, penetrating, and moving away from the surface. With the excavation surface close to the surface, the settlement rate shows a trend of increasing and then decreasing. For tunnels with upper and lower multiple guide tunnels, the construction of the upper guide tunnels has a greater impact on the surface settlement. For tunnels with multiple guide tunnels on the left, center, and right, the surface in the middle is affected by the overlap of the construction of the left and right guide tunnels, which results in a greater settlement than the left and right surfaces.

(2) For the surrounding rock stress law, the initial stress state of the tunnel around the points of the surrounding rock is in a state of pressure, and the size of the stress is positively correlated with the depth of burial. After the support structure is applied, the deformation of the surrounding rock gradually decreases until the surrounding rock and the support structure reach a new equilibrium. After the removal of the diaphragm wall that provides vertical support and the temporary elevated arch that provides horizontal support, the connection between the initial support and the temporary support shrinks and deforms inwardly, and the value of the stress decreases and transfers to the surroundings.

(3) Lining the stress law, the stress value of the support structure increases because of the force given by the surrounding rock after the excavation of the guide hole. After the excavation of the lower guide hole and application of the support structure, the stress value of the lower structure increases while the stress value of the upper structure decreases. After the excavation of the middle and upper guide holes and the application of the support structure, the stress value of the two sides of the structure increases. After the middle and lower guide holes are excavated and the supporting structure is applied, the lining stress of the whole structure is redistributed and concentrated downward. When the temporary support is removed, the stress on the diaphragm wall and the temporarily elevated arch is transferred to be borne by the initial support, and the stress value of the initial support increases.

(4) Compared with the upper and lower excavation method, the surface settlement caused by the left and right excavation method is smaller, and the timely formation of a closed loop of the support structure has a certain control effect on the surface settlement. The left and right excavation method causes less disturbance to the surrounding rock, which indicates that the overall structure can better restrain the surrounding rock. The support structure is allowed to be more evenly and stably loaded during the excavation process, and the excavation disturbances have a smaller impact on the support structure.

Author Contributions: Conceptualization, Y.R. and Y.L.; methodology, Y.R.; validation, X.L., and J.L.; formal analysis, Y.L. and B.Y.; investigation, Y.R. and Y.L.; resources, X.L. and B.Y.; data curation, S.L. and C.L.; writing—original draft preparation, Y.R.; writing—review and editing, Y.L. and B.Y.; supervision, C.L. All authors have read and agreed to the published version of the manuscript.

Funding: National Natural Science Foundation of China: No. 51978177 and 52278336; Guangdong Basic and Applied Research Foundation: No. 2023B1515020061 and 2022A1515240037; Research and Development Project of The Ministry of Housing and Urban Rural Development (2022-k-044).

Data Availability Statement: The data presented in this study are available on request from the corresponding author. The data are not publicly available due to privacy.

Conflicts of Interest: Authors Yanmei Ruan, Xu Luo and Shan Lin were employed by the company Guangzhou Metro Design & Research Institute Co., Ltd. Authors Jin Li and Chengkun Ling were employed by the company China Railway Tenth Bureau Group Urban Rail Transit Engineering Co., Ltd. The remaining authors declare that the research was conducted in the absence of any commercial or financial relationships that could be construed as a potential conflict of interest.

References

1. Liu, J.Y.; Lv, G.; Zhang, M.Q.; Yue, L.; Luo, D.H. Research on Excavation Methods of Extra-Large Span Tunnel at Badaling Great Wall Station of Jingzhang High-Speed Railway. *Mod. Tunneling Technol.* **2019**, *56*, 578–584.
2. Liu, C. Analysis of the Influence of Different Excavation Methods on the Stability of Highway Tunnels. *Highw. Eng.* **2019**, *44*, 130–134.
3. Yang, J. Comparison of Construction Methods for Tunnels Adjacent to the Runway Section of Huanghua Airport in Changhai High-Speed Railway. *Mod. Tunneling Technol.* **2021**, *58*, 118–126.
4. Zhao, M.; Cheng, Y.; Song, Z.; Wang, T.; Zhang, Y.; Liu, L.B. Optimization of Construction Parameters and Deformation Characteristics of Large-Section Loess Tunnel: A Case Study from Xi'an Metro. *Adv. Civ. Eng.* **2021**, *2021*, 6639089. [CrossRef]
5. Bai, B.; Nie, Q.K.; Zhang, Y.K.; Wang, X.L.; Hu, W. Cotransport of heavy metals and SiO_2 particles at different temperatures by seepage. *J. Hydrol.* **2021**, *597*, 125771. [CrossRef]
6. Yuan, B.X.; Liang, J.K.; Lin, H.Z.; Wang, W.Y.; Xiao, Y. Experimental Study on Influencing Factors Associated with a New Tunnel Waterproofing for Improved Impermeability. *J. Test. Eval.* **2023**, *52*, 20230417. [CrossRef]
7. Liu, X.B.; He, S.H.; Wang, D.H.; Zhang, J.; Yao, W.B. Mechanical Characteristics of Initial Support System for Shallow-Buried Super-Large Span Four-Line High-Speed Railway Tunnels During Construction. *China Railw. Sci.* **2021**, *42*, 90–102.
8. Liu, D.P.; Zhang, D.L.; Fang, Q.; Sun, Z.Y.; Luo, J.W.; Li, A. Field Monitoring of the Deformation and Internal Forces of the Surrounding Rock and Support Structures in the Construction of a Super-Span High-Speed Railway Tunnel-A Case Study. *Appl. Sci.* **2020**, *10*, 5182. [CrossRef]
9. Zhao, Y.; Li, P.F. Performance of a multi-face tunnel excavated in loess ground based on field monitoring and numerical modeling. *Arab. J. Geosci.* **2016**, *9*, 640.
10. Bai, B.; Rao, D.Y.; Chang, T.; Guo, Z.G. A nonlinear attachment-detachment model with adsorption hysteresis for suspension-colloidal transport in porous media. *J. Hydrol.* **2019**, *578*, 124080. [CrossRef]
11. Wei, H.H.; Cui, G.Y. Study on Underground Excavation Construction Technology of TBM with Tunnel First Excavated and then Expanded into Platform Based on Double Side Drift Method. *Geotech. Geol. Eng.* **2022**, *40*, 3595–3605.
12. Chen, Y.; Tang, L.S.; Sun, Y.L.; Cheng, Z.H.; Gong, W.T. Physical-mechanical properties and microstructure degradation of acid-alkali contaminated granite residual soil. *Geomech. Energy Environ.* **2023**, *36*, 100501. [CrossRef]
13. Cui, C.Y.; Liang, Z.M.; Xu, C.S.; Xin, Y.; Wang, B.L. Analytical solution for horizontal vibration of end-bearing single pile in radially heterogeneous saturated soil. *Appl. Math. Model.* **2023**, *116*, 65–83. [CrossRef]
14. Yuan, B.X.; Li, Z.H.; Zhao, Z.Q.; Ni, H.; Su, Z.L.; Li, Z.J. Experimental study of displacement field of layered soils surrounding laterally loaded pile based on Transparent Soil. *J. Soils Sediments* **2021**, *21*, 3072–3083. [CrossRef]
15. Zhu, G.Q.; Feng, X.T.; Fu, Y.u. Real-time monitoring of the development of brittle fracture in hard rock tunnels based on physical model test. *Tunn. Undergr. Space Technol.* **2022**, *119*, 104240. [CrossRef]
16. Que, C.H.; Zhu, Z.D.; He, Y.X.; Niu, Z.H.; Huang, H.N. Strength and deformation characteristics of irregular columnar jointed rock mass: A combined experimental and theoretical study. *J. Rock Mech. Geotech. Eng.* **2023**, *51*, 429–441. [CrossRef]
17. Wang, C.; Su, Z.M.; Zhao, J. Study on Optimization of Core Pillar in Large-span Tunnel Excavated with Double-side Heading Method. In Proceedings of the International Conference on Civil Engineering and Building Materials (CEBM), Kunming, China, 29–31 July 2011; p. 1882.
18. Bai, B.; Zhou, R.; Cai, G.Q.; Hu, W.; Yang, G.H. Coupled thermo-hydro-mechanical mechanism in view of the soil particle rearrangement of granular thermodynamics. *Comput. Geotech.* **2021**, *137*, 104272. [CrossRef]
19. Peng, F.J.; Zhang, X.; Li, X.Z.; Jiang, B.; Liu, B.H.; Zhang, H.J. Optimization Analysis of Construction Scheme for Large-Span Highway Tunnel Under Complex Conditions. *Arch. Civ. Eng.* **2018**, *64*, 55–68.

20. Que, X.C.; Zhu, Z.D.; Zhou, L.M.; Niu, Z.H.; Huang, H.N. Strength and failure characteristics of an irregular columnar jointed rock mass under polyaxial stress conditions. *Rock Mech. Rock Eng.* **2022**, *55*, 7223–7842. [CrossRef]
21. Yuan, B.X.; Chen, W.J.; Li, Z.H.; Zhao, J.; Luo, Q.Z.; Chen, W.W.; Chen, T.Y. Sustainability of the polymer SH reinforced recycled granite residual soil: Properties, physicochemical mechanism, and applications. *J. Soils Sediments* **2023**, *23*, 246–262. [CrossRef]
22. Que, X.C.; Zhu, Z.D.; Niu, Z.H.; Zhu, S.; Wang, L.X. A modified three-dimensional Hoek–Brown criterion for intact rocks and jointed rock masses. *Geomech. Geophys. Geo-Energy Geo-Resour.* **2023**, *9*, 7. [CrossRef]
23. Yang, Z.H.; He, R.M.; Li, K.; Guo, H.Y. Parameter optimization of double side wall method for soft surrounding rock tunnel. *E3S Web Conf.* **2019**, *136*, 04021. [CrossRef]
24. Yuan, B.X.; Chen, W.J.; Zhao, J.; Li, L.J.; Liu, F.; Guo, Y.C.; Zhang, B.F. Addition of alkaline solutions and fibers for the reinforcement of kaolinite-containing granite residual soil. *Appl. Clay Sci.* **2022**, *228*, 106644. [CrossRef]
25. Li, X.D.; Wang, W. Construction Technology of Reserving Core Rock Pillars in Shallow-Buried Tunnel with Double-Sided Wall Guideway Method for Metro. *Tunn. Constr.* **2015**, *35*, 1060–1065.
26. Zhang, L.L.; Wu, F.; Wei, X.; Yang, H.Q.; Fu, S.X.; Huang, J.S.; Gao, L. Polynomial chaos surrogate and bayesian learning for coupled hydro-mechanical behavior of soil slope. *Rock Mech. Bull.* **2023**, *2*, 100023. [CrossRef]
27. Jiang, Y.J.; Li, B.; Wang, C.S.; Song, Z.Y.; Yan, B.M. Advances in development of shear-flow testing apparatuses and methods for rock fractures: A review. *Rock Mech. Bull.* **2022**, *1*, 100005. [CrossRef]
28. Meng, k.; Cui, C.Y.; Liang, Z.M. A new approach for longitudinal vibration of a large-diameter floating pipe pile in visco-elastic soil considering the three-dimensional wave effects. *Comput. Geotech.* **2020**, *128*, 103840. [CrossRef]
29. Chen, Y.; Tang, L.S.; Ye, Y.H.; Cheng, Z.H.; Zhou, Z.X. Effects of different chloride salts on granite residual soil: Properties and water–soil chemical interaction mechanisms. *J. Soils Sediments* **2023**, *4*, 1844–1856. [CrossRef]
30. Yang, B.B.; Chen, Y.; Zhao, C.; Li, Z.L. Effect of geotextiles with different masses per unit area on water loss and cracking under bottom water loss soil conditions. *Geotext. Geomembr.* **2023**. [CrossRef]
31. Li, X.Z. Experimental study on surface settlement model of extremely shallow buried large-span multi-arch tunnels. *J. Wuhan Univ. Technol. (Transp. Sci. Eng. Ed.)* **2012**, *36*, 1118–1121.

applied sciences

Article

Influence of Bedding Strength and Angle on Fracture Characteristics of Sandstone under Three-Point Bending Conditions †

Guobo Wang [1], Tao Meng [1,2,*], Gan Feng [3], Lifeng Ma [1], Weimin Yang [1], Yi He [1], Zhijiang Zhang [1] and Xufeng Liang [1]

1 School of Mechanical Engineering, Taiyuan University of Science and Technology, Taiyuan 030024, China
2 Mining Technology Institute, Taiyuan University of Technology, Taiyuan 030024, China
3 State Key Laboratory of Hydraulics and Mountain River Engineering, College of Water Resource & Hydropower, Sichuan University, Chengdu 610065, China
* Correspondence: mengtao@tyust.edu.cn; Tel.: +86-15513053025

† The research results of this paper can be used as the theoretical basis for further development and application of coal in situ gasification technology.

Abstract: The fracture characteristics of bedded sandstone determine the stability and safety of in situ coal gasification technology. Four semicircular stratified sandstone specimens with different strengths (0.3, 0.6, 1.0, and 1.5 times that of rock matrix) and seven different bedding angles ($\theta = 0°$, 15°, 30°, 45°, 60°, 75°, and 90°) were numerically simulated using RFPA2D-Basic V2.0 software. The SCB specimen had no prefabricated crack, and its radius was 25 mm. The loading rate was 0.000001 m/step. The results show that the fracture characteristics of the sandstone are affected by both the strength of the laminae and the angle; the fracture toughness and peak strength of the ultra-weak sandstone, as well as the weak sandstone, are reduced and more easily affected by the bedding angle; the strength rate of the strong sandstone is higher than that of the homogeneous sandstone, and the difference between the fracture characteristics of the two is not significant. This paper suggests that the key mechanism of this phenomenon is the anisotropy between the bedding and the sandstone, along with the competition/synergy between the main crack and the bedding plane bias crack during fracture propagation. These research results can provide a theoretical reference for the safety and stability of underground engineering in China.

Keywords: sandstone; bedding strength; bedding angle; three-point bending; fracture properties; failure mode

check for updates

Citation: Wang, G.; Meng, T.; Feng, G.; Ma, L.; Yang, W.; He, Y.; Zhang, Z.; Liang, X. Influence of Bedding Strength and Angle on Fracture Characteristics of Sandstone under Three-Point Bending Conditions. *Appl. Sci.* **2023**, *13*, 8216. https://doi.org/10.3390/app13148216

Academic Editor: Bing Bai

Received: 23 May 2023
Revised: 7 July 2023
Accepted: 11 July 2023
Published: 15 July 2023

1. Introduction

Lignite accounts for a large proportion of China's coal reserves. According to the national coal resource survey, the proven reserves of brown coal are 131.142 billion tons. This is equivalent to 13% of China's total coal reserves (1.00326 trillion tons) [1]. However, due to its high oxygen content, lignite is prone to spontaneous combustion. Lignite has a high water content and is difficult to store. In addition, due to the high contents of volatile substances and relatively low heat content in lignite, burning lignite for power generation often leads to serious air pollution. Therefore, China's research focus is on the efficient and clean utilization of lignite resources. Underground in situ pyrolysis and gasification is an important method to enable clean utilization of lignite in China. Some scholars have found that one of the most effective methods is underground pyrolysis and gasification. The pyrolysis and gasification of coal refers to heating coal to high temperatures without air and producing pyrolysis products through physical and chemical changes, such as gases (CH_4, H_2, and CO_2), liquids (tar), and solids (coke) [2,3]. In the process of underground coal pyrolysis and gasification, gas needs to seep into the outlet pipeline through the surrounding coal and rock mass, but if the gas cannot smoothly penetrate into the outlet pipeline, it may diffuse and penetrate into other surrounding rocks, causing environmental pollution [4]. In the process of coal pyrolysis and gasification, the movement of gas is mainly controlled by the permeability and mechanical properties of the surrounding coal and rock mass.

The fracture properties of layered rock mass, as shown in Figure 1, are divided into two main parts: mechanical properties, and crack propagation behavior [5]. Among them, the mechanical properties of layered rock have been studied by scholars. Li et al., (2007) conducted direct shear tests on three specimens of layered salt rock without an interlayer, pure hard gypsum, and laminated salt rock with a hard gypsum interlayer, and they obtained the shear strength parameters of the three specimens, finding that the shear strength parameters at the intersection of the hard gypsum interlayer and salt rock were slightly stronger than those of the pure laminated salt rock and pure hard gypsum specimens [6]. To investigate the effects of different test conditions on shale crack extension, Fan et al., (2008) considered the laminar dip angle, the height-to-width ratio of the specimen, and the rate of test loading, and the elastic modulus and compressive strength of the shale both decreased and then increased with the increasing angle between the laminar surface and the loading direction, while the length of the main crack increased and then decreased [7]. Nasseri et al., (1996) conducted uniaxial and triaxial tests on four types of schist and analyzed the variation of strength and deformation anisotropy with the dip angle and confining pressure in detail [8]. Wasantha et al., (2014) studied the peak strength and deformation characteristics of layered sandstone with different bedding dip angles in dry and saturated states, and they analyzed the energy dissipation characteristics of the sample under loading deformation and failure by using the monitored acoustic emission data [9]. Yao et al., (2018) found that the mechanical properties of layered sandstone under static loading showed significant differentiation with the change in the dip angle [10].

(a) (b)

Figure 1. (**a**) Layered red sandstone with vertical, horizontal, and inclined fractures through bedding contacts (Badami, India). (**b**) The folded layers in a valley in Namibia with potential sites for vertical, horizontal, and inclined interface cracks [11].

Some scholars have also studied the crack propagation behavior of layered rock mass, e.g., Heng et al., (2014) analyzed the influence of the bedding direction on the expansion behavior of hydraulic cracks [12]. The fracture propagation behavior of layered rock mass has also been studied by scholars; for example, using layered phyllite, Li et al., (2015) investigated the deformation and failure of layered rock mass under dynamic splitting tensile strain and discovered that there are three different failure mechanisms, depending on the bedding angle [13]. Based on the three-point bending test and Brazilian disc splitting test, Heng et al., (2015) studied the evolution mechanism of crack propagation in shale samples with well-bedding. Under the condition of three-point bending, when the bedding intersects horizontally or diagonally, the crack path will be offset along the bedding plane, and the smaller the bedding angle, the stronger the crack propagation ability will be. When the bedding is vertical, the crack will propagate along the bedding plane until the failure is complete and the fracture plane is smooth and flat. The crack propagation under the Brazilian splitting condition also has a similar phenomenon to that mentioned above [14]. Xiao (2017) studied fracture propagation behavior in sand–mudstone interbedded rock mass based on the finite element method and found that the difference between bedding strength and principal stress has a significant influence on the fracture propagation and fracturing pressure [15]. Huang et al., (2020) conducted experimental and numerical three-point bending tests on semicircular sandstone specimens with open-mode cracks to investigate the influence of bedding parameters on their mechanical fracture behavior. The experimental study focused on the effects of the loading direction and laminar dip angle on the fracture loading, fracture morphology, and damage process [16].

However, while the angle of layers, layer thickness, and layer spacing are key criteria impacting the construction, the strength qualities of the layered surface (e.g., degree of cementation) are equally important [17,18]. As a result, the author used RFPA2D-Basic software to create a numerical model of semicircular (SCB) layered sandstone specimens, and then conducted a numerical simulation study of semicircular bending with seven different laminar dips for laminated sandstone specimens containing four different strength layers of cells (without preset fractures), combined with the stress–strain curve, peak stress, acoustic emission (AE) location map (Pang et al., 2014; Zuo et al., 2019; Liu et al., 2021; Xue et al., 2021) [19–22], AE energy (or the number of events), and damage cloud map of the model. The mechanical properties and damage patterns of the laminated sandstone specimens under semicircular bending loading conditions were investigated, and the failure damage mechanism was analyzed by studying the entire process of crack initiation, extension evolution, and penetration of the sandstone specimens. This provides a theoretical reference for the safety and stability of artificial storage lanes in geothermal reservoirs in China.

2. Model Establishment and Parameter Selection

2.1. Model Building

The geometric characteristics of the numerical model and the loading method are shown in Figure 2.

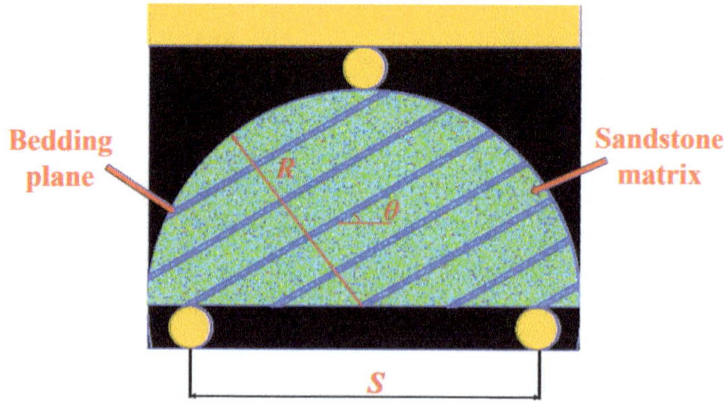

Figure 2. The model of layered sandstone.

The height of the whole grid cell model was 40 mm, the width was 50 mm, and the grid cell was divided into 280 × 224 = 62,720 square-shaped meso-elements. The radius of the semicircular specimen model R = 25 mm; three circular rigid bodies were established as supports for the three-point bending experiment, with a radius of 2.5 mm, and the span ratio of the bottom support was taken as 0.8 (span S = 40 mm), while a bearing plate was also established at the top with the same parameters as the supports. In this paper, the angle θ between the laminar surface direction and the horizontal direction is defined as the laminar surface inclination angle, and a total of seven different laminar surface inclination angles (θ = 0°, 15°, 30°, 45°, 60°, 75°, and 90°) are set in this paper [23], with a laminar surface thickness of 0.6 mm and a layer spacing of 4.46 mm. As the bedding density remains constant, the number of beddings varies with the bedding angle in the semicircular sample model. For example, when θ = 0°, there are four bedding layers in the sample model, and when θ = 90°, there are nine bedding layers in the sample model. According to the experience of using the software, to make the simulation effect better and the calculation results more accurate, we carried out optimization of the modeling [24]. During the modeling, the mouse was used to draw the layering directly on the grid interface, instead of establishing the layering by inputting coordinate parameters. In this simulation, the loading method of displacement control in the standard model was used. The bottom of the model was fixed, and the pressure was applied on the bearing plate with the initial value set to 0.000001 m, after which each step was loaded with the same 0.000001 m until the specimen was damaged.

2.2. Parameter Selection

The macro-mechanical parameters used in this simulation of sandstone are referred to as the data measured in the uniaxial compression experiments in Peng's paper [25], as shown in Table 1. The RFPA2D Basic V2.0 software introduced the Weibull distribution function to characterize the non-uniformity of the properties of the medium, with the homogeneity coefficient m as the link to establish the medium's macro-mechanical parameters and mesoscopic mechanical parameters. The fine mechanical parameters of the medium can be calculated by the following fitting equation:

$$\begin{aligned} \frac{E_{macroscopic}}{E_{microscomic}} &= 0.1412 \ln m + 0.6476 (1.2 \leq m \leq 10) \\ \frac{f_{macroscopic}}{f_{microscomic}} &= 0.2602 \ln m + 0.0233 (1.2 \leq m \leq 50) \end{aligned} \quad (1)$$

where $E_{macroscopic}$ represents the macroscopic strength of the dielectric material, $f_{macroscopic}$ represents the macroscopic elastic modulus of the dielectric material, $E_{microscopic}$ represents the mesoscopic mean of the dielectric strength, and $f_{microscopic}$ represents the mesoscopic mean of the dielectric elastic modulus.

Table 1. Macroscopic mechanical parameters.

Compressive Strength (MPa)	Elasticity Modulus (GPa)	Poisson Ratio	Internal Friction Angle (°)	Density (kg/m^3)
57.5	12.7	0.15	40	2400

The homogeneity coefficient m of strength and elastic modulus is 2, and the homogeneity of Poisson's ratio and density is 100. The calculated meso-mechanical parameters and other input parameters of the model are summarized in Table 2.

Table 2. Model input parameters.

Category of the Parameter	Name of Parameter	Value of the Parameter	Name of Parameter	Value of the Parameter
Control parameters	Residual strength/MPa	0.1	Internal friction angle/°	40
	Residual Poisson's ratio	1.1	Tension and compression ratio	10
	Failure criterion	Mohr–Coulomb	Maximum compressive strain	Control variable
	Maximum tensile strain	1.5	Single-step load/mm	0.001

In this simulation study, four different bedding planes were set, and the strength of the bedding plane was taken as 0.3, 0.6, 1.0, and 1.5 times the strength of the sandstone matrix, while all other parameters were kept constant. It should be noted that the laminae with 1.0 times the strength of the matrix are homogeneous sandstone, and they were set as the control group to more intuitively reveal the effect of bedding strength on the fracture characteristics of sandstone. Due to the large coordinate parameters required to draw the bedding plane, the vertex coordinates of each bedding plane are listed in Table 3 only with the inclination angle of the bedding plane being 45°. The bedding planes are numbered from left to right.

Table 3. Coordinate parameters of the bedding planes.

Number of Bedding Plane	Left Vertex Coordinates	Right Vertex Coordinates
1	(0.0004, 0.0094) (0.0002, 0.0084)	(0.0206, 0.0296) (0.0216, 0.0298)
2	(0.0031, 0.0050) (0.0040, 0.0050)	(0.0279, 0.0298) (0.0287, 0.0297)
3	(0.0103, 0.0050) (0.0111, 0.0050)	(0.0337, 0.0284) (0.0343, 0.0282)
4	(0.0174, 0.0050) (0.0183, 0.0050)	(0.0385, 0.0261 (0.0390, 0.0257)
5	(0.0246, 0.0050) (0.0254, 0.0050)	(0.0425, 0.0229) (0.0429, 0.0225)
6	(0.0317, 0.0050) (0.0326, 0.0050)	(0.0457, 0.0190) (0.0461, 0.0185)
7	(0.0389, 0.0050) (0.0397, 0.0050)	(0.0482, 0.0143) (0.0484, 0.0137)
8	(0.0460, 0.0050) (0.0469, 0.0050)	(0.0497, 0.0087) (0.0498, 0.0079)

3. Analysis of the Tensile Strength

3.1. Stress–Strain Curve

The longitudinal stress–strain curves of sandstone samples with different bedding angles and strengths were drawn as shown in Figure 3. It can be seen from the figure that the longitudinal stress–strain curves of the sandstone samples in each group have similar changing trends, which have experienced an elastic deformation stage, yield stage, brittle failure stage, and residual failure stage. The curves are all straight lines at the beginning, and the stress and strain are proportional, in contrast to the experimental results of previous scholars. This is because there is a compaction stage in the conventional experimental compression process, and the compression structure of sandstone samples is denser under the action of the pressure head, so the curve is an arc curve rather than a straight line. With the continuous loading of displacement, the slope of the curve decreases somewhat, but it is not obvious. When the peak value is reached, the curve drops sharply. Brittle failure occurs in the sandstone sample at this time, and then it enters the residual failure stage, at which point the whole failure process ends. The slope of the curve decreases with continuous displacement loading, but it is not very obvious; when the peak is reached, the curve drops steeply; obviously, the sandstone specimen suffers brittle damage at this time, and then after entering the residual damage phase, to the end of the entire damage process.

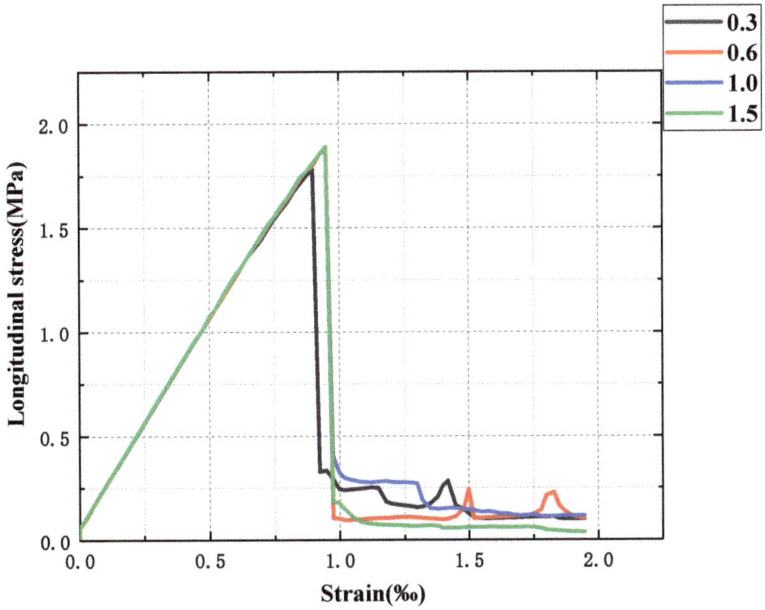

(**a**)

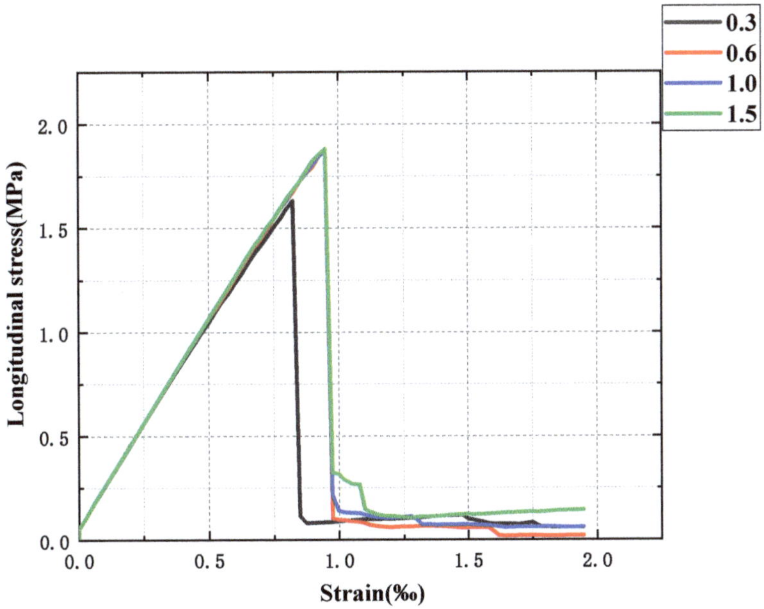

(**b**)

Figure 3. *Cont.*

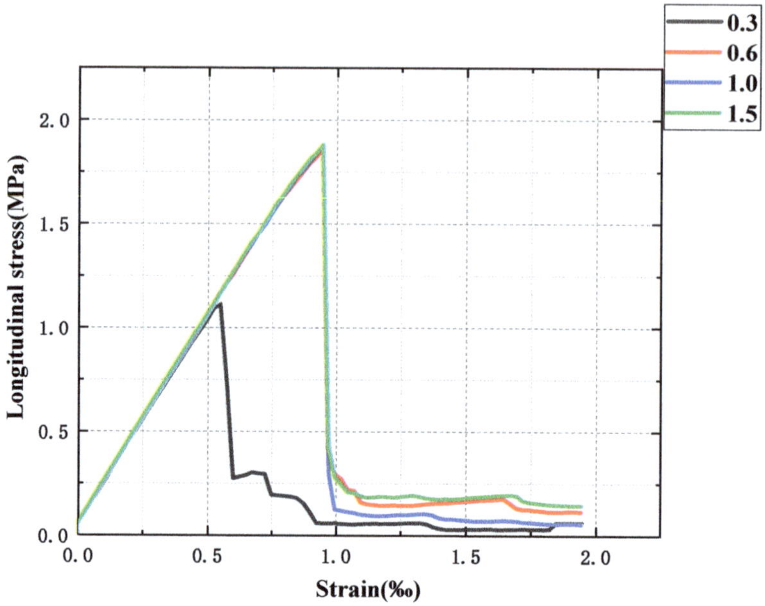

(c)

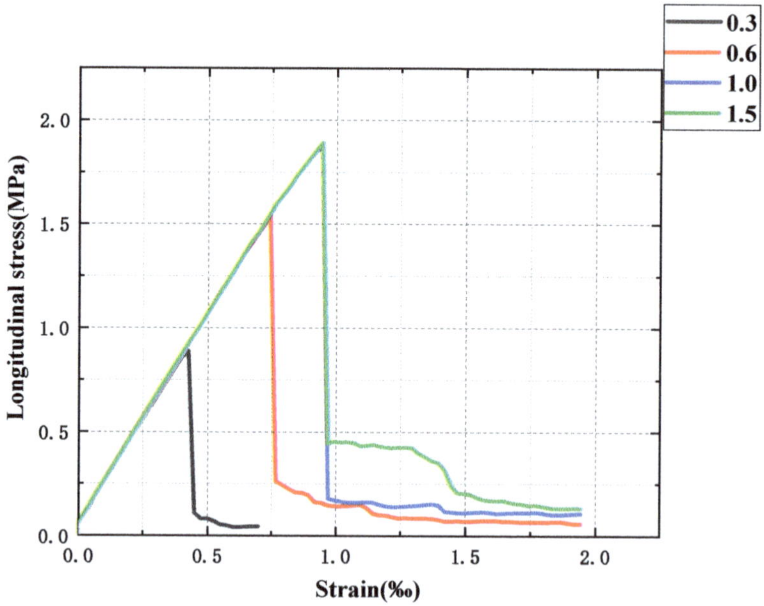

(d)

Figure 3. *Cont.*

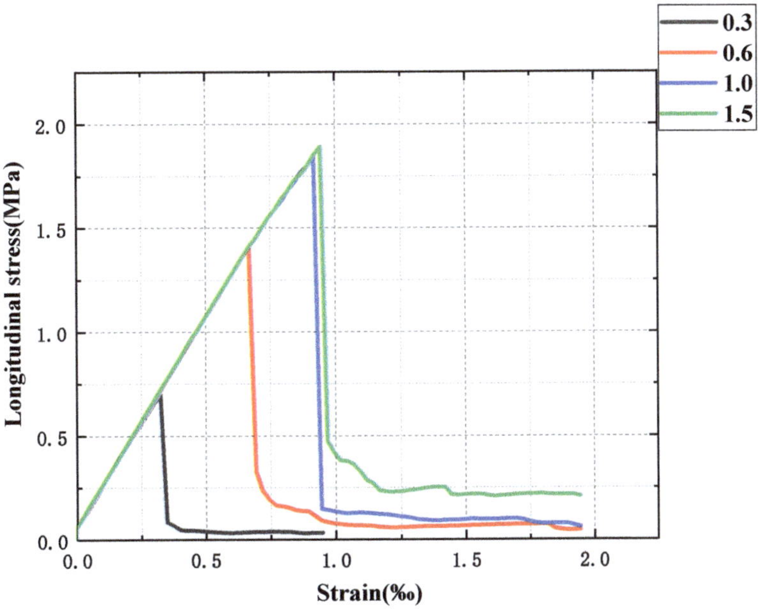

(e)

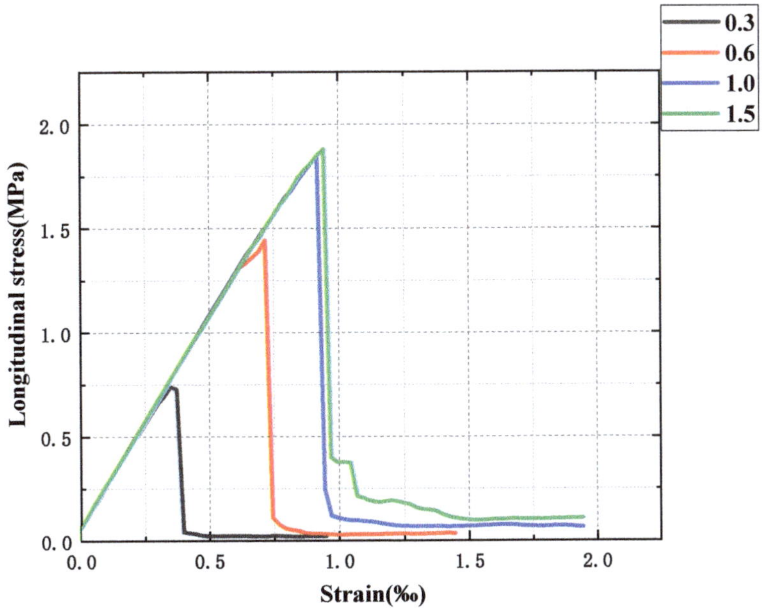

(f)

Figure 3. *Cont.*

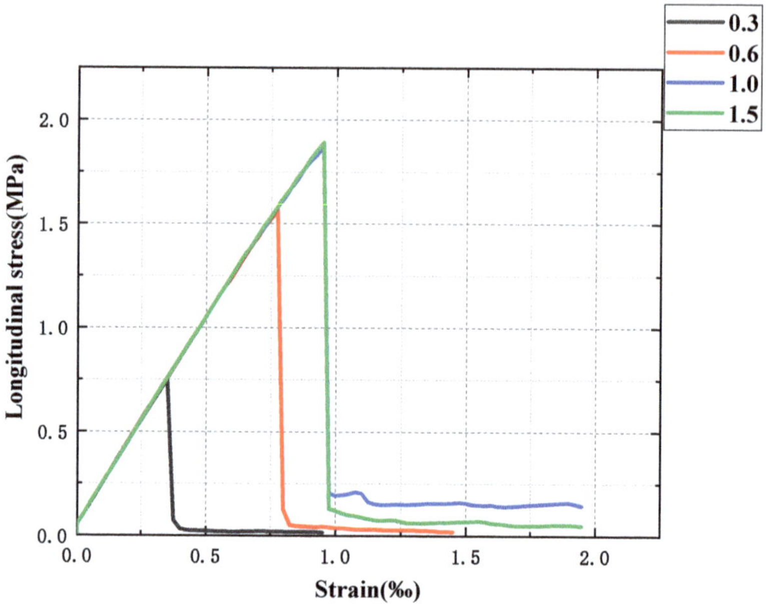

(g)

Figure 3. Stress-strain curves of different sandstone. (**a**) 0°; (**b**) 15°; (**c**) 30°; (**d**) 45°; (**e**) 60°; (**f**) 75°; (**g**) 90°.

By comparing the stress–strain curves of ultra-weak bedding (bedding strength of 0.3 times the sandstone strength), weak bedding (bedding strength of 0.6 times the sandstone strength), strong bedding (bedding strength of 1.5 times the sandstone strength), and homogeneous sandstone (bedding strength of 1 times sandstone strength), we found the following: (1) With the change in the bedding angle, the strength of the ultra-weak sandstone decreases obviously. However, the variation is not linear; from 0° to 15°, the variation range is small, but when the bedding angle is greater than 30°, the strength of the ultra-weak sandstone decreases sharply, to only half of the strength of homogeneous sandstone, and reaches the minimum value at 60°, which is about 1/3 of the strength of the homogeneous sandstone. The observation of the strain value of sandstone when the fracture occurs shows that the strain of homogeneous sandstone is about 1% when the fracture occurs, while that of the ultra-weak sandstone is 0.85% when the bedding angle is 0° and drops to 0.6% when the bedding angle is 30°, and then decreases slowly with the increase in the angle until it reaches the minimum value at 60°; when the strain is less than 0.4%, the sandstone has fractured. (2) With the increase in the angle of the bedding, the change in the strength of the weakly bedded sandstone is relatively small, and there is an obvious weakening of the strength only after 45°, and the weakening is significantly smaller than that of the ultra-weak bedded sandstone; the strength of the weakly bedded sandstone decreases by 20% compared with that of the homogeneous sandstone, but it is also far beyond that of the ultra-weakly bedded sandstone, at about 1.5–2 times the strength of the latter. At the same time, the change law of strain and the strength of different layer angles are essentially the same. (3) Strong sandstone (bedding strength of 1.5 times that of sandstone) has no strength difference from homogeneous sandstone, which is somewhat stronger than the mean strength of sandstone. In addition, by comparing the curves of sandstone with different bedding angles, it was found that the strength of all sandstone with bedding strength is the highest at 0° and the lowest at 60°. With the increase in the bedding angle, the sandstone strength shows a "U"-type change pattern; that is, it first decreases, reaches the minimum value at 60 h, and then increases.

3.2. Peak Stress

According to the above stress–strain curves, the peak stress of sandstone with different bedding angles and different bedding strengths was extracted. It can be seen from Figure 4 that the peak stress of sandstone is controlled by the bedding angle θ and the bedding strength when the bedding strength is smaller than the matrix. With the increasing bedding dip angle, the peak stress of ultra-weak sandstone and weak sandstone first declines and then climbs, forming an asymmetric "U" shape as a whole, with the smallest value at 60°. Peak stress exhibits a distinct decreasing pattern from 15° to 60° and a sluggish increasing trend from 60° to 90°; when the sandstone's bedding strength surpasses its strength, the bedding angle has no effect on the peak stress.

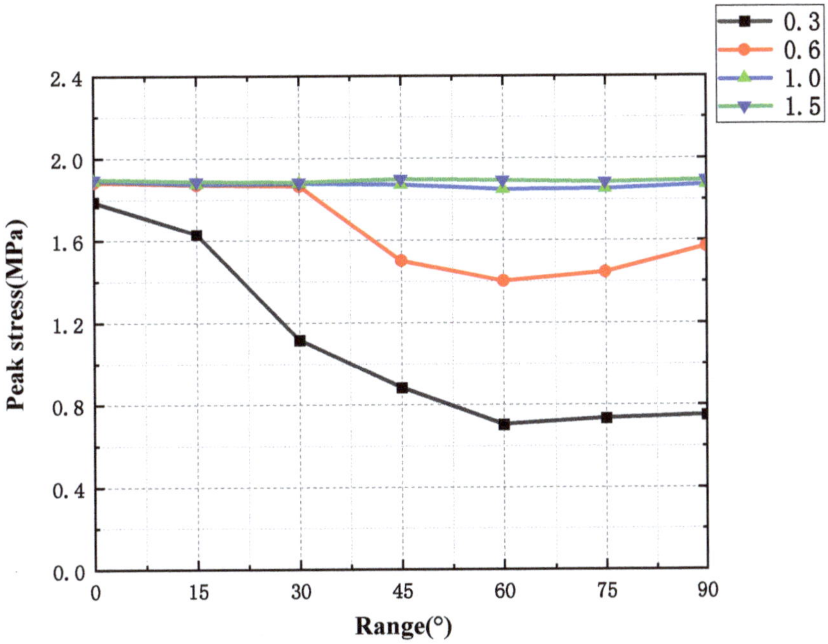

Figure 4. Peak stress curves of different sandstone specimens.

The reason for this phenomenon is that the bedding planes of the above two specimens are both weak, and their existence will weaken the bearing capacity of the rock mass and make its failure stress level lower than that of the intact rock mass (i.e., the strength of the bedding and the matrix is equal). When θ = 0°and 15°, the damage model of the rock mass is the tensile damage that occurs through the bedding and along the bedding rupture, and the bearing capacity of the rock mass at this time mainly depends on the strength of the rock matrix, so its peak is slightly smaller than that of the intact rock mass. When θ = 30°, the damage type is mainly tension damage along with the bedding; local penetration rupture is produced, and its bearing capacity is affected by the joint influence of bedding strength and matrix strength. When θ = 60°, the damage model is tension damage completely along with the bedding, and its load-bearing capacity depends entirely on the strength of the bedding surface, so the peak value is the lowest. When θ = 45°, 75°, or 90°, the rock body occurs completely along with the bedding tension damage, and the bearing capacity mainly depends on the bedding strength, while the degree of influence of its matrix strength on the bearing capacity varies with the change in the dip angle. In addition, the interaction between cracks can also affect its carrying capacity, and the peak stress is larger than when θ = 60. When the strength of the bedding reaches the strength of the matrix, the peak stress does not affect the bedding angle. When the strength of the bedding increases within a certain range, the peak stress does not change with the strength of bedding. This is because even if the bedding strength is increased, the thickness of the bedding surface is very thin compared to the matrix, and the properties of the two are relatively similar at this time; the rock mass is a homogeneous material, and the mechanical properties are isotropic, so the bearing capacity will not be affected by anything.

4. Analysis of Damage Characteristics

In order to investigate the damage mechanism of sandstone with different bedding angles and strengths, the damage process and damage mode were studied in two aspects by combining the stress cloud map obtained from the simulations and the acoustic emission results. The various stages of the damage process of layered sandstone were analyzed in depth, and the types of damage modes of laminated sandstone under the influence of lamina strength and dip angle were summarized.

4.1. Damage Process

It is mentioned in Section 3.1 that the failure process of layered sandstone goes through the elastic deformation stage, yield stage, brittle damage stage, and residual failure stage. In this section, the acoustic emission curve and stress cloud map are combined to further study this process. It should be noted that, according to the previous results, we found that although the data on the strength and peak stress of sandstone with different bedding strengths and different bedding angles were different, the damage process was similar. Therefore, this section studies weakly layered sandstone, and the results can be applied to other sandstone.

In Figure 5, each stage of the damage process is analyzed:

1. Elastic deformation stage: Stress increases linearly with strain, and the slope of the curve remains constant. With the continuous loading of the displacement, the strain energy gradually accumulates, but the number of AE events does not change significantly, and no cracks are generated at this time.
2. Yield stage: This is also the stage of crack emergence, expansion, and penetration. When the stress exceeds 50% of the peak value, the slope of the curve decreases somewhat, but it is not obvious; there is a trace of AE events, but no obvious fluctuations; at this time, the tiny cracks sprout. When the stress exceeds 80% of the peak value, the number of AE events fluctuates slightly, but there is no obvious increase, and the micro-cracks begin to expand and penetrate to form larger cracks. A small amount of plastic deformation occurs in the rock at this stage, but it is not significant.
3. Brittle damage stage: After the stress exceeds the peak value, the curve drops suddenly, the accumulated strain energy is released instantly, and the number of AE events shows a steep increase in the number of single peaks; the cracks generated in the yield stage extend rapidly, and the specimen undergoes brittle damage.
4. Residual damage stage: The brittle damage occurs in an instant, followed by the residual damage stage, where the AE signal suddenly disappears. The stress level is extremely low in this stage, and the specimen is biased. In the whole damage process, the elastic stage is the longest, followed by the yield stage. The brittle damage stage is instantaneous, so the residual damage stage, when the rock body has destroyed the end, strictly speaking, cannot be attributed to the main process of damage. The figure shows a schematic diagram of the stress bar in the y-directional stress field damage cloud, in which the negative sign represents the compressive stress and the opposite is the tensile stress. It can be seen that the y-directional stress field damage cloud varies with the stress–strain curve as follows: the stress state before the peak stage is compressive.

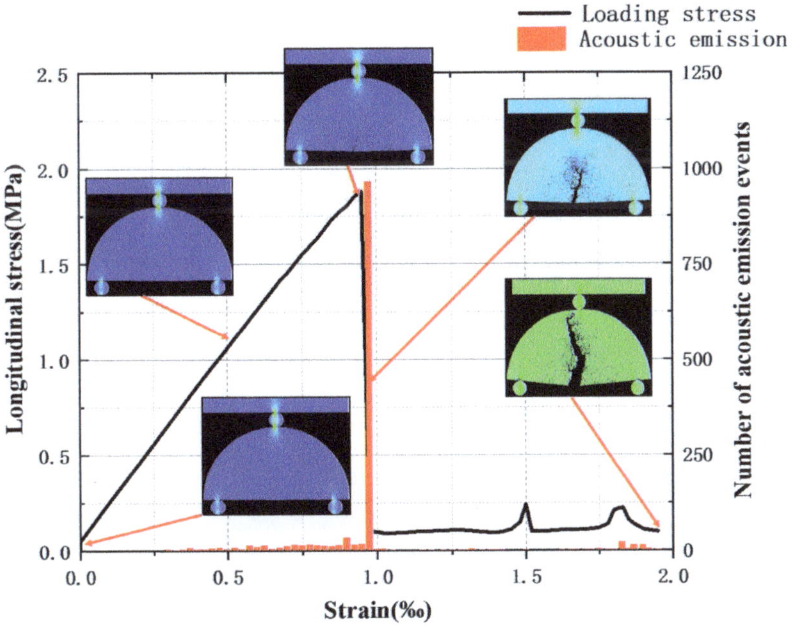

(**a**)

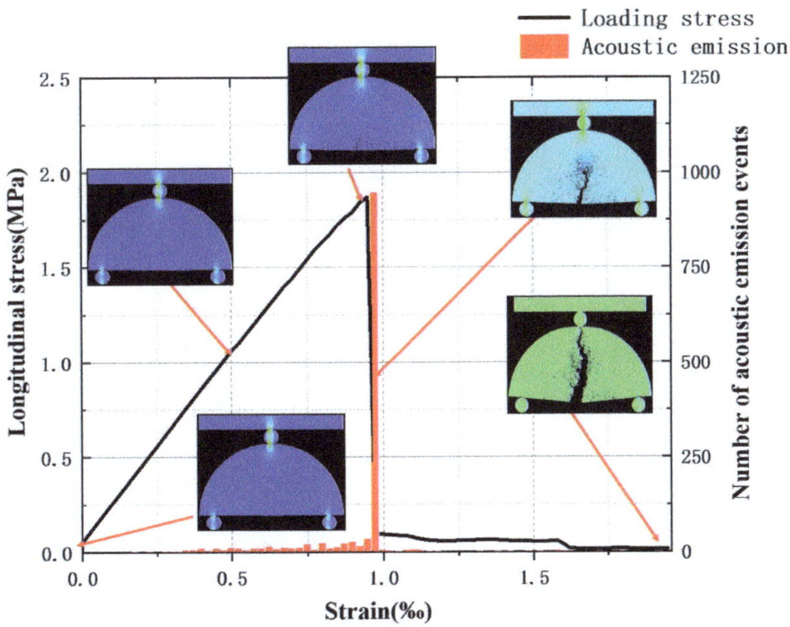

(**b**)

Figure 5. *Cont.*

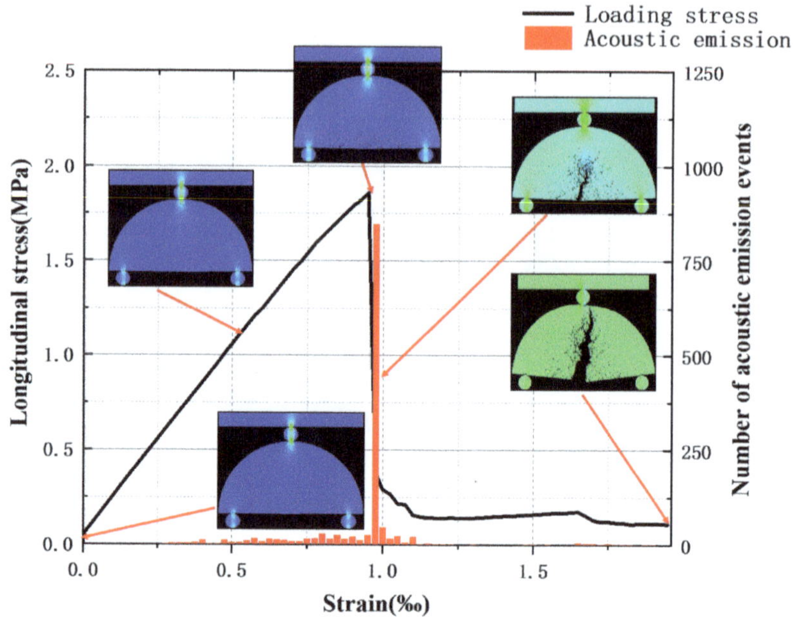

(**c**)

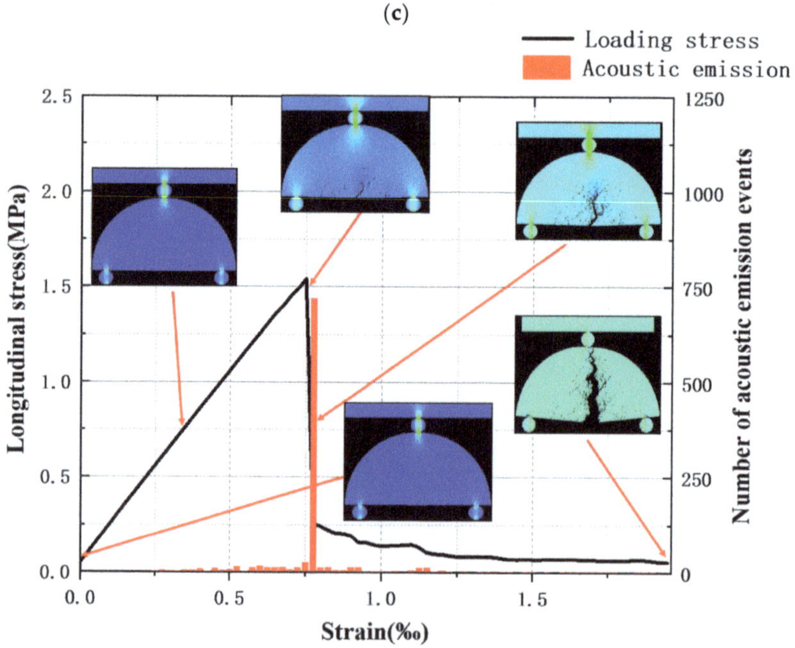

(**d**)

Figure 5. *Cont.*

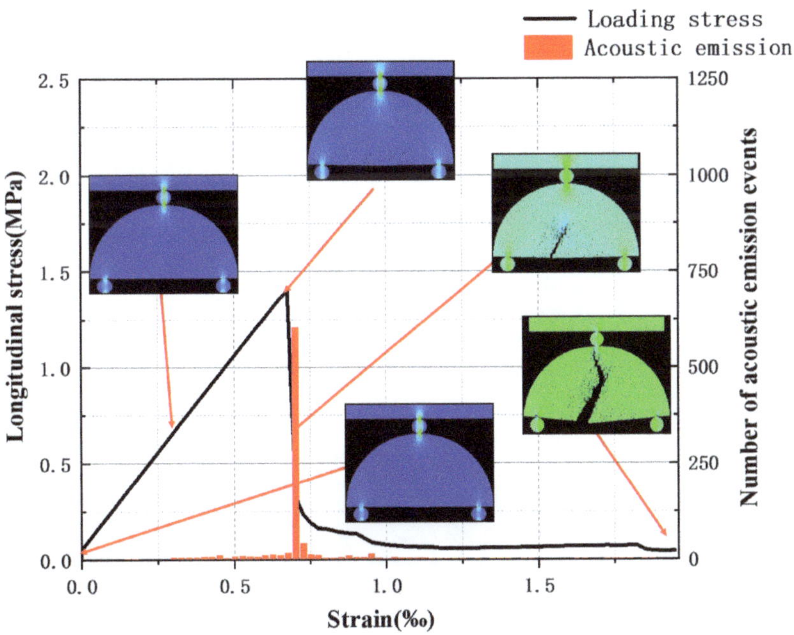

(e)

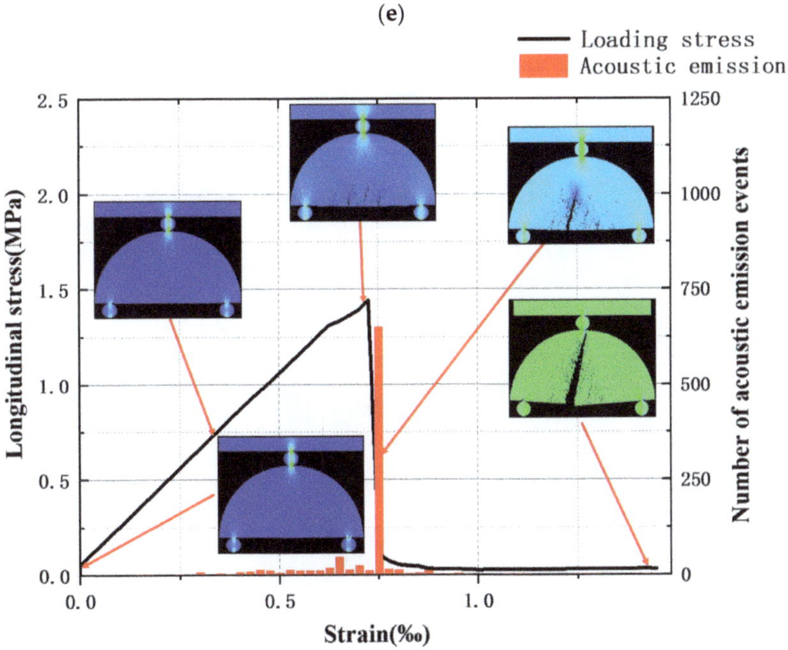

(f)

Figure 5. *Cont.*

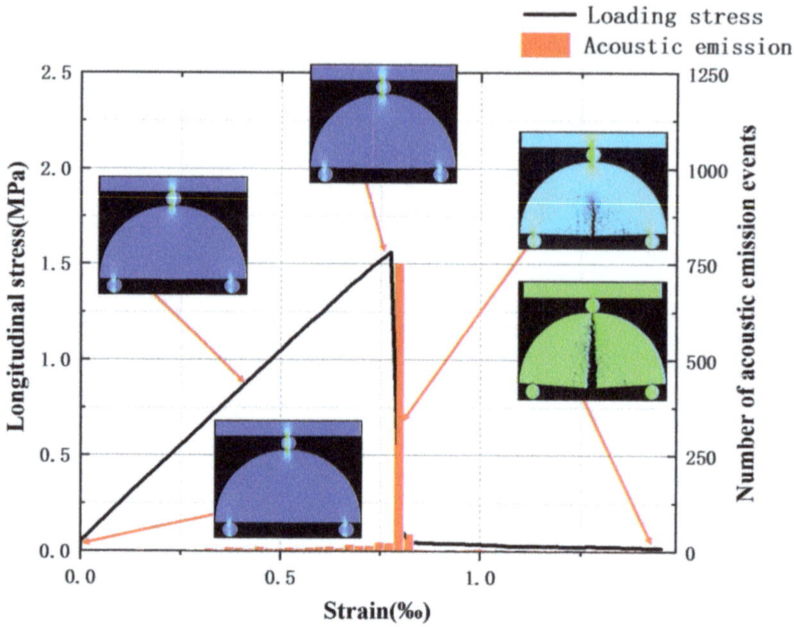

(g)

Figure 5. Fracture processes of different sandstone: **(a)** 0°; **(b)** 15°; **(c)** 30°; **(d)** 45°; **(e)** 60°; **(f)** 75°; **(g)** 90°.

4.2. Destruction Mode

From the damage process, it can be seen that different bedding angles lead to different crack propagation trajectories. By compiling the shear stress damage clouds of the residual stages of the four bedding strength specimens under different bedding angles and focusing on the process of crack propagation (Figure 6), the damage patterns of each group of specimens were analyzed as follows:

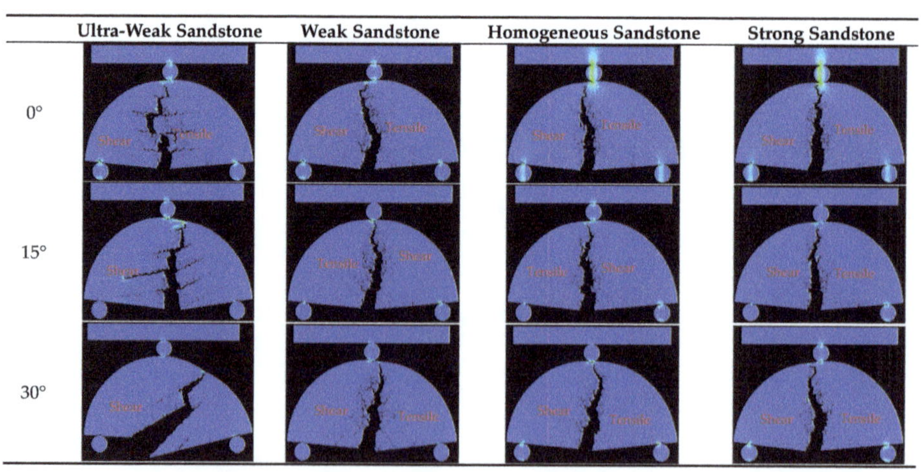

Figure 6. *Cont.*

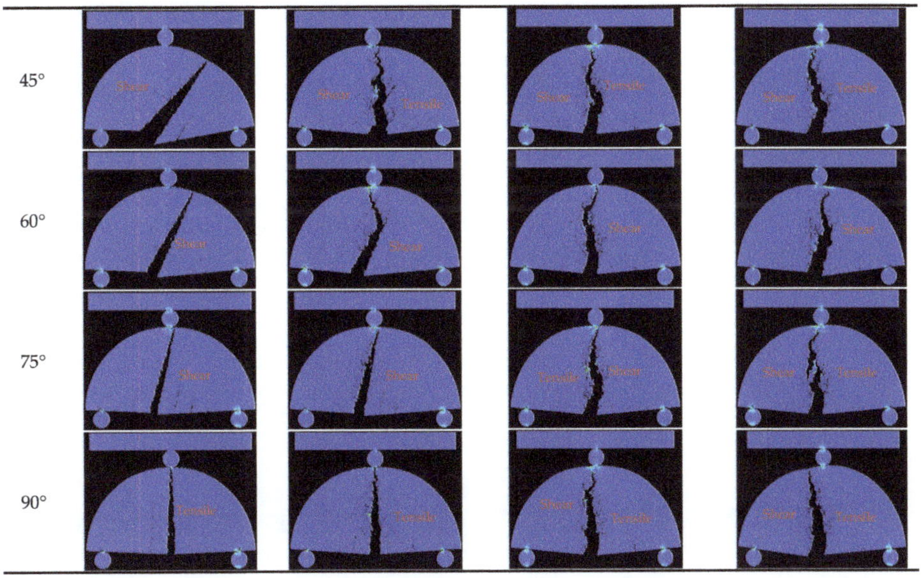

Figure 6. Cloud images of different sandstone fracture modes.

For the ultra-weakly bedded sandstone, the rupture patterns were similar at $\theta = 0°$ and 15°. Advanced cracks occurred on the bedding plane, and multiple secondary cracks were generated along the bedding plane. The primary cracks started from the middle matrix part at the bottom of the specimen and expanded toward the loading point, but the secondary cracks led the primary cracks to migrate along the bedding surface, and the crack expansion path eventually showed an intermittent step pattern. Therefore, the rupture mode was a mixed type of tension rupture with mainly through-layer tension rupture, accompanied by the migration of the main crack along the bedding surface. When $\theta = 30°, 45°, 60°, 75°$, or 90°, the layered surface will also rupture superficially, and multiple cracks along the layered surface will appear. With the loading of displacement, these cracks interact with one another; that is, the cracks in the rock body fracture competitively with one another, and eventually the crack at a certain layer "win" becomes the main crack and expands until the end of the damage. When $\theta = 30°$, the crack expands mainly along the 30° bedding plane, but localized penetration rupture occurs in the process. Therefore, the damage mode is dominated by the tension rupture along the bedding surface, and local tension rupture through the layers is produced. At other angles, the cracks expand along the bedding plane until the end of the damage, so the entire damage mode is tension rupture along the laminae.

In the case of weakly bedded sandstone, the damage patterns are similar for $\theta = 0°$, 15°, 30°, and 45°. Advanced cracks still occur on the bedding plane, but they are not as obvious as when the bedding strength is 0.3 times, and the guiding effect of secondary cracks on the main crack is obviously weakened. The damage model is still a mixed type of tension rupture with mainly tension rupture through the plies, accompanied by the migration of the main crack along the bedding plane. With the increase in the bedding inclination angle, the extension direction of the main crack tends to be more vertical. When $\theta = 60°$, the damage mode is mainly tension rupture along the bedding plane, and the transbedding tensile fracture occurs locally. When $\theta = 75°$ and 90°, the crack completely develops a tensile fracture along the bedding plane.

In the case of the strongly bedded sandstone, the damage pattern of each laminated dip specimen is very similar to that of the homogeneous sandstone, which shows that the strength and dip angle of the bedding have little effect on the damage pattern of the specimens at this time, and the damage patterns that occur are essentially the same, with all of them being tension rupture through the layers in the direction of the loading point. Among them, when the bedding angle is 30° and 60°, there is a great difference between the fracture trajectory of homogeneous sandstone and the fracture trajectory of the two in the form of "symmetry". These two bedding angles lead to relatively active offset fractures and frequent shear fractures, leading to a large offset of the main fractures, similar to ultra-weak sandstone. The peak stresses during the fracture of sandstone with different bedding strengths all reach the minimum value at the bedding angle of 60°, which also indicates that the offset cracks show a certain activity at this time. A portion of the stress is applied along the direction of the bedding plane, causing a shear fracture to occur and, thus, the crack deflection phenomenon.

5. Discussion

Combining the contents of Sections 3 and 4, we believe that there are two main reasons for the different fracture characteristics of sandstone with different bedding angles and different bedding strengths: Anisotropy between the bedding and the sandstone, along with the competition/synergy between the main cracks and the bedding surface, offsets cracks during fracture expansion. The sandstone fracture characteristics of different bedding strengths are mainly because of the former, and the sandstone fracture characteristics of different bedding angles are mainly because of the latter. The effects of the lamina angle and the bedding strength on the fracture characteristics of sandstone are discussed in detail in both directions below.

5.1. Lamination and Sandstone Anisotropy

By comparing the fracture modes of the four sandstone specimens, we found that the crack propagation trajectories of ultra-weak sandstone and homogeneous sandstone are quite different, which we attributed to the large anisotropy between the bedding strength and the collective strength of the sandstone. Based on the stress–strain curve, we can see that the strength of ultra-weak bedding is worse than that of homogeneous sandstone. According to the peak stress curve and Formulas (2) and (3), it can be concluded that the fracture toughness of ultra-weak sandstone is also relatively poor.

$$K_{IC} = Y' \frac{P_{\max}\sqrt{\pi\alpha}}{2RB} \tag{2}$$

$$Y' = -1.297 + 9.516(S/R) - [0.47 + 16.457(S/R)]\beta + [1.071 + 34.401(S/R)]\beta^2 \tag{3}$$

where K_{IC} is the fracture toughness, P_{max} is the fracture load, Y' is the critical dimensionless factor derived from the finite element method, and $\beta = \alpha/R$. The strength and the fracture toughness increase as the strength of the laminae increases; therefore, we believe that the strength and the fracture toughness of the ultra-weakly laminated sandstone bedding are small and cause the overall strength and fracture toughness to be affected. When the ultra-weakly laminated sandstone receives external load through divine tension damage, it is relatively more prone to fracture due to the low laminar fracture toughness. Due to the presence of the bedding and the adhesive force between it and the sandstone matrix, the fractured bedding structure will exert some tension on the sandstone matrix in the horizontal direction, which will cause the sandstone matrix to fracture under a much smaller load. The difference between the bedding strength and fracture toughness of weakly bedded sandstone and those of the sandstone matrix is small, so the overall strength and fracture toughness of weakly bedded sandstone will decrease, but the decrease is small. As for the strongly bedded sandstone, because its bedding strength and fracture toughness are greater than those of the sandstone matrix, when the sandstone matrix fractures, its bedding has not yet fractured, so some compressive force will be applied to the sandstone matrix in the horizontal direction. Therefore, the peak stress of strongly stratified sandstone will be slightly higher than that of homogeneous sandstone.

In addition, when the crack propagates unsteadily in the vertical layer of the layered material, the laminar peeling phenomenon will occur due to the relatively weak inter-laminar bonding force. We can imagine the laminated sandstone as a whole consisting of several individual sandstone layers as well as a combination of laminae, and each sandstone layer is subjected to different tensile forces when the cracks expand. In the case where the strength of the bedding plane, as well as the fracture toughness, is less than that of the sandstone matrix, when the first sandstone layer cracks, the second sandstone layer is subjected to a greater tensile force than the first, because the bedding plane exerts a tensile force on the second part of the sandstone, and so on—the individual sandstone layers are subjected to increasing tensile forces, so the laminated sandstone in general will fracture under smaller loads. Conversely, when the strength and fracture toughness of the bedding surface are smaller than those of the sandstone matrix, the individual sandstone layers are subjected to less and less tensile force due to the compressive force exerted by the bedding on the sandstone matrix, so the laminated sandstone as a whole needs a greater load to fracture.

5.2. Competition/Synergy between the Main Crack and the Layer Surface Offset Crack

By comparing the fracture extension trajectories of ultra-weakly laminated sandstone with different lamina angles, we found that they were more different from those of homogeneous sandstone. It is obvious from the fracture mode diagram that the interference of the bedding plane during the crack extension—especially at 0°–30°—causes the main crack of the fracture to be severely disturbed by the bedding plane. This is due to the inter-laminar slip effect. When the bedding intersects the loading direction obliquely at a small angle ($\theta < 45°$), the sandstone specimens from the interfacial shear slip at the bedding plane under the axial load. As there is less adhesion between the bedding plane and the sandstone matrix, shear slip causes the bedding plane to separate from the sandstone matrix, resulting in the appearance of offset cracks along the bedding plane. When the bedding angle is small, the offset cracks favor the horizontal direction and the main cracks favor the vertical direction, and the competition between them leads to a large variation in the crack expansion trajectory. When the bedding angle is large, the offset cracks are biased in the vertical direction and the main crack is also in the vertical direction, and the two do not compete with one another, but instead promote one another, so the crack expansion trajectory does not change much.

6. Conclusions

The deformation damage characteristic was evaluated using the finite element program RFPA2D Basic V2.0, based on three-point bending tests of semicircular beam (SCB) specimens, and considering the layer inclination and strength. The following are the key findings of this study:

Under the influence of the weak surface and dip angle of laminae, the mechanical properties of sandstone have significant anisotropy. When the lamina strength is 0.3 and 0.6 times that of the rock matrix, the peak stress tends to decrease and then increase with the increase in the lamina dip angle, the overall shape is similar to a "U", and $\theta = 60°$ is the cutoff point, i.e., the peak stress is the smallest at this time; under the same lamina dip angle, the peak stress gradually increases with the increase in the lamina.

The failure mode of sandstone is controlled by the joint control of the weak surface of the lamina and the dip angle. When $\theta = 0°$ and 15°, the weak plane will show advanced fracture along the bedding direction, and the secondary cracks will lead the main crack to migrate along the bedding direction and then hinder the main crack from expanding in the direction of the loading.

When $\theta = 30°$, 45°, 60°, 75°, or 90°, several cracks appear on the weak surface along the bedding direction, and one of the cracks finally competes successfully and begins to expand. When $\theta = 30°$, the cracks partially break through the bedding, while at other angles the cracks continue to expand along the bedding direction until the end of failure.

The number of AE events corresponds to the failure mode, and the distribution range of AE events is the widest when the sample has a mixed fracture of cross-layer fracture and the fracture along the bedding plane. The distribution range of AE events is the smallest when the fracture is completely along the bedding plane. When pure translaminar rupture occurs, AE events have a small distribution range.

When the strength of the bedding is weaker than that of the rock matrix, the cumulative number of AE events at $\theta = 0°$ and $15°$ gradually decreases with the strengthening of the bedding. The cumulative number of AE events at $\theta = 30°$ first increases and then decreases, while the cumulative number of AE events at the remaining dip angles gradually increases, and when the strength of the laminae reaches that of the rock matrix, the cumulative number of AE events at all dip angles tends to be equal.

There was no preset crack in this study, and the failure type was not affected by the bedding strength and dip angle. According to the AE location maps of each sample, it can be seen that the failure type was a tensile failure. By observing the whole process of crack propagation, the failure mode of the samples can be divided into the following four types: pure lamellar tensile failure; dominant translaminar tensile fracture, accompanied by the migration of the main crack along the bedding plane; dominant tensile fractures along the bedding, with transbedding tensile fractures occurring locally; and complete tensile fracture along the bedding.

Author Contributions: Conceptualization, G.W. and T.M.; methodology, T.M., G.W. and G.F.; investigation, G.W. and T.M.; resources, L.M.; software, W.Y.; formal analysis, Z.Z.; writing—original draft preparation, Z.Z.; writing—review and editing, Y.H.; project administration, X.L. All authors have read and agreed to the published version of the manuscript.

Funding: This research was funded by the financial support from the Youth Program of the National Natural Science Foundation of China (51904195), the Scientific and Technological Innovation Programs of Higher Education Institutions in Shanxi (2019L0640), the Key Research and Development Project of Shanxi Province (201903D121025), the Fundamental Research Program of Shanxi Province (201901D211300; 20191142), TYUST-SRIF (20172018), and the Outstanding Doctoral Award Fund in Shanxi Province (20192032). Experimental data can be found at the NGDC repository of the British Geological Survey (ID 136076). We also wish to thank the National Natural Science Foundation of China (Grant No. 52104143); The Fundamental Research Funds for the Central Universities (Grant No. 2021SCU12039); and the Natural Science Foundation of Sichuan Province, China (Grant No. 2022NSFSC0193).

Data Availability Statement: No new data were created or analyzed in this study. Data sharing is not applicable to this article.

Conflicts of Interest: The authors declare no conflict of interest.

References

1. Yin, L.Q. Lignite resources and utilization outlook in China. *Coal Sci. Technol.* **2004**, *8*, 12–14. (In Chinese)
2. Anthony, D.B.; Howard, J.B. Coal devolatilization and hydrogastification. *AIChE J.* **1976**, *22*, 625–656. [CrossRef]
3. Arenillas, A.; Rubiera, F.; Pis, J.J.; Cuesta, M.; Iglesias, M.; Jiménez, A.; Suárez-Ruiz, I. Thermal behaviour during the pyrolysis of low rank perhydrous coals. *J. Anal. Appl. Pyrolysis* **2003**, *68–69*, 371–385. [CrossRef]
4. Humenick, M.J.; Mattox, C.F. Groundwater pollutants from underground coal gasification. *Water Res.* **1978**, *12*, 463–469. [CrossRef]
5. Tian, Z.; Tang, C.; Liu, Y.; Tang, Y. Zonal disintegration test of deep tunnel under plane strain conditions. *Int. J. Coal Sci. Technol.* **2020**, *7*, 337–349. [CrossRef]
6. Li, Y.P.; Jiang, W.D.; Liu, J.; Chen, J.W.; Yang, C.H. Direct shear test of deep layered salt rock in Yunying Salt Mine, Hubei Province. *J. Rock Mech. Eng.* **2018**, *26*, 6.
7. Fan, X.Y.; Guo, D.Y.; Zhang, Q.G.; Xu, F.L.; Liang, Y.C.; Lu, X.W. Experimental study on shale crack propagation mechanism considering bedding, loading rate and size effects. *Sci. Technol. Eng.* **2018**, *18*, 9.
8. Nath, F.; Mokhtari, M. Optical visualization of strain development and fracture propagation in laminated rocks. *J. Pet. Sci. Eng.* **2018**, *167*, 354–365. [CrossRef]
9. Wasantha, P.L.P.; Ranjith, P.G.; Shao, S.S. Energy monitoring and analysis during deformation of bedded-sandstone: Use of acoustic emission. *Ultrasonics* **2014**, *54*, 217–226. [CrossRef]

10. Yao, N.; Ye, Y.C.; Wang, Q.H.; Luo, B.Y.; Wang, W.Q. Numerical simulation of uniaxial compression behavior of gently inclined layered rock mass by particle flow. *J. Eng. Geol.* **2018**, *26*, 9.
11. Alneasan, M.; Behnia, M.; Bagherpour, R. Analytical investigations of interface crack growth between two dissimilar rock layers under compression and tension. *Eng. Geol.* **2019**, *259*, 105188. [CrossRef]
12. Heng, S.; Yang, C.H.; Cao, Y.J.; Guo, T.Y.; Wang, L.; Hou, Z.K. Experimental study on fracture morphology of shale hydraulic fracturing. *J. Geotech. Eng.* **2014**, *36*, 9.
13. Li, Q. The view of technological innovation in coal industry under the vision of carbon neutralization. *Int. J. Coal Sci. Technol.* **2021**, *8*, 1197–1207. [CrossRef]
14. Heng, S.; Yang, C.H.; Guo, T.Y.; Wang, C.Y.; Wang, L. Effect of bedding on hydraulic fracture propagation in shale. *J. Rock Mech. Eng.* **2015**, *34*, 228–237.
15. Xiao, W. Propagation of fracture in thin interbedded sand and mudstone. *J. Chengde Pet. Coll.* **2017**, *19*, 4.
16. Huang, D.; Li, B.; Ma, W.-Z.; Cen, D.-F.; Song, Y.-X. Effects of bedding planes on fracture behavior of sandstone under semi-circular bending test. *Theor. Appl. Fract. Mech.* **2020**, *108*, 102625. [CrossRef]
17. Kawakami, G.; Kawamura, M. Sediment flow and deformation (SFD) layers: Evidence for intrastratal flow in laminated muddy sediments of the Triassic Osawa Formation, northeast Japan. *J. Sediment. Res.* **2002**, *72*, 171–181. [CrossRef]
18. He, J.; Afolagboye, L.O. Influence of layer orientation and interlayer bonding force on the mechanical behavior of shale under Brazilian test conditions. *Acta Mech. Sin.* **2018**, *34*, 349–358. [CrossRef]
19. Pang, Y.; Wang, G.; Ding, Z. Mechanical model of water inrush from coal seam floor based on triaxial seepage experiments. *Int. J. Coal Sci. Technol.* **2014**, *1*, 428–433. [CrossRef]
20. Zuo, J.; Wang, J.; Jiang, Y. Macro/meso failure behavior of surrounding rock in deep roadway and its control technology. *Int. J. Coal Sci. Technol.* **2019**, *6*, 301–319. [CrossRef]
21. Liu, B.; Zhao, Y.; Zhang, C.; Zhou, J.; Li, Y.; Sun, Z. Characteristic strength and acoustic emission properties of weakly cemented sandstone at different depths under uniaxial compression. *Int. J. Coal Sci. Technol.* **2021**, *8*, 1288–1301. [CrossRef]
22. Xue, D.; Lu, L.; Zhou, J.; Lu, L.; Liu, Y. Cluster modeling of the short-range correlation of acoustically emitted scattering signals. *Int. J. Coal Sci. Technol.* **2021**, *8*, 575–589. [CrossRef]
23. Fu, J.; Haeri, H.; Sarfarazi, V.; Naderi, A.A.; Marji, M.F.; Xu, L. Influence of arch shaped notch angle, length and opening on the failure mechanism of rock like material and acoustic emission properties: Experimental test and numerical simulation. *Theor. Appl. Fract. Mech.* **2023**, *125*, 103879. [CrossRef]
24. Sarfarazi, V.; Naderi, A.; Zhou, L.; Haeri, H.; Fatehimarji, M. Study on shear fracture behavior of soft filling in concrete specimens: Experimental tests and numerical simulation. *Struct. Eng. Mech.* **2023**, *85*, 337–351.
25. Peng, Y.; Liu, G.; Wang, C.Y.; Wang, D.C.; Jia, P. Effect of inhomogeneity on macroscopic creep characteristics of sandstone. *Sichuan Build. Mater.* **2021**, *47*, 82–83, 85.

applied sciences

MDPI

Article

Influence of the Construction of the Urdinbide Road Tunnel on the Autzagane Aquifer in Biscay (Spain)

José-Manuel Baraibar [1,*], Miguel Gil [2] and Iñigo Escobal [1]

[1] Viuda de Sainz, S.A., Technical and Innovation Department, P. El Campillo 19, 48500 Abanto-Zierbena, Spain; iescobal@viudadesainz.com

[2] Interbiak, Kanariar Uharten Kalea, 19, 48015 Bilbao, Spain; mgil@interbiak.eus

* Correspondence: jmbaraibar@viudadesainz.com

Abstract: The Urdinbide road tunnel goes through the Autzagane aquifer. This important aquifer is located within the hydrogeological area of the Urdaibai Biosphere Reserve, one of the most important biosphere reserves in the Iberian Peninsula, and it is also used as a source of drinking water for some urban areas in the municipality of Amorebieta-Etxano. The construction of the tunnel could pose a potential risk to the normal functioning of the aquifer, so its design included a special procedure for injecting cement and microcement to waterproof the surrounding area of the drilling, preventing the tunnel from functioning as a drain for the aquifer. The project initially included an intensive hydrogeological characterization of the rock massif, as well as a monitoring phase during the construction works, which allowed a real-time verification of the influence of the tunnel construction on the functioning of the aquifer and the restoration of its normal functioning once the works were completed. The work carried out has shown that the construction of the tunnel has not caused a significant impact on the Autzagane aquifer.

Keywords: tunnel; multilayer aquifer; underground construction; cement; microcement; injections; Flysch; waterproofing

Citation: Baraibar, J.-M.; Gil, M.; Escobal, I. Influence of the Construction of the Urdinbide Road Tunnel on the Autzagane Aquifer in Biscay (Spain). *Appl. Sci.* **2023**, *13*, 7034. https://doi.org/10.3390/app13127034

Academic Editor: Bing Bai

Received: 13 May 2023
Revised: 4 June 2023
Accepted: 8 June 2023
Published: 11 June 2023

1. Introduction

The Urdinbide road tunnel is part of a new road communication route between Amorebieta and Muxika (Figure 1). It is in the northern part of the Iberian Peninsula, specifically in the Basque Country, running through the municipalities of Amorebieta-Etxano and Muxika in the province of Bizkaia (Spain).

Figure 1. Location of the study area. Relative situation with respect to the metropolitan area of Bilbao, capital of Biscay, in the north of Spain.

The environmental approval of the project was conditioned on the execution of the tunnel not significantly affecting the Autzagane aquifer, which it was going to pass through, as well as maintaining the existing water catchment and usage systems and the ecological flow of the Ategorri stream, which comes from the springs that supply this aquifer.

During the design phase, the hydrogeological functioning of the aquifer was studied in depth. In addition, during its construction, a thorough hydrogeological monitoring was carried out, including the monitoring of piezometric levels and flow rates. The data obtained were used to calibrate the initial hydrogeological simulation of the aquifer. This hydrogeological monitoring made it possible to quantify the effectiveness of the proposed waterproofing solution both during the construction phase and in the final phase, once the tunnel was completed and waterproofed [1].

There are many references in the scientific literature that analyze the internal dynamics of complex aquifers [2–14] and the influence of the construction of underground infrastructures in these highly permeable environments [15–18] and that study possible strategies to mitigate this phenomenon [19–21].

The Urdinbide tunnel project is characterized by its proximity to the Urdaibai area, of high ecological value, which was declared a biosphere reserve by UNESCO in 1984 [22]. In addition, it has the particularity that the Autzagane aquifer is used to supplement the water supply of an important population center in the municipality of Amorebieta-Etxano. To prevent water from entering the tunnel and draining the aquifer, a strategy of rock mass injection was employed during the construction of the tunnel, which in this particular case consisted of a combination of cement and microcement in stages. In addition, a completely impermeable lining was installed to ensure that the aquifer was not affected in service conditions.

In the particular case of the Urdinbide tunnel, systematic, repetitive, and selective injections (IRS type) were used, with a high-pressure approach [23]. These injections were carried out using a redrilling system in the same boreholes, with incremental lengths in each drilling phase. These were truncated cone-shaped drilling and grouting crowns executed in a minimum set of four stages with cement grouting in the first stages and microcement grouting in the last stages. This set of injections also served to improve the geotechnical quality of the rock mass around the tunnel.

The tunnel waterproofing system should ensure the recovery of the piezometric levels and the functioning of the streams in a normal rainfall regime, as well as maintaining the original compartmentalization of the aquifer and its natural hydrogeological functioning, as imposed by the environmental authorization of the project. The main purpose of this contribution is to describe the influence of the Urdinbide tunnel waterproofing solution on the hydrogeological functioning of the Autzagane aquifer.

2. Materials and Methods

2.1. General Description of Tunnel Works

The Urdinbide tunnel is the main element of the new road between Amorebieta and Muxika. This new road replaces a heavily trafficked route with a high accident rate [1]. It is safer and improves accessibility from the A8 motorway to the Urdaibai Biosphere Reserve area.

The Urdinbide tunnel passes through the Autzagane Pass. It starts at KP 2 + 683, near the entrance to the Biribieta neighborhood and at an altitude of approximately +185 m. It ends at KP 3 + 386, at an altitude of around +180 m, adopting a straight route with a descending slope towards Muxika. The Urdinbide tunnel is twin-tube, with two lanes per carriageway. The average separation between the two tubes is 21.50 m, practically constant throughout all the tunnel.

The length of the tunnel is around 700 m with maximum coverings of around 90 m (Figure 2). This length includes the 116 and 56 m of false tunnel at the south and north portals, respectively.

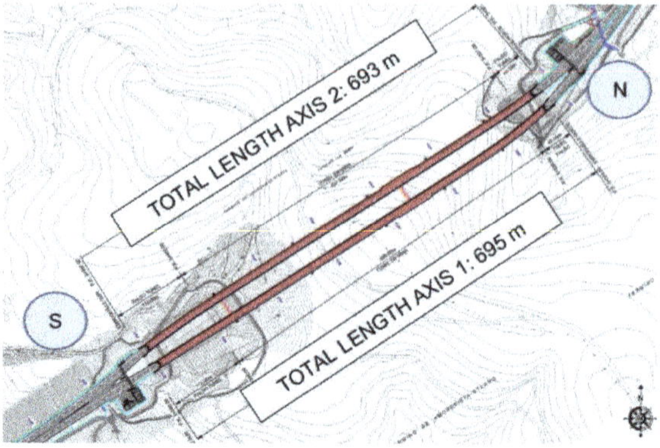

Figure 2. General layout of the Urdinbide tunnel.

The affected area of the aquifer by the tunnel is limited between KPs 2 + 859 and 3 + 019 on Axis 1 (160 m) and between KPs 2 + 840 and 3 + 004 on Axis 2 (164 m). The Urdinbide tunnel has a circular section with an inner radius of 6.53 m in the area outside the aquifer and 6.48 m in the aquifer area (Figure 3), with a height at the center of 1.286 m above the axis with the road surface along the whole length of the tunnel, achieving a minimum clearance of 5 m above the platform edges.

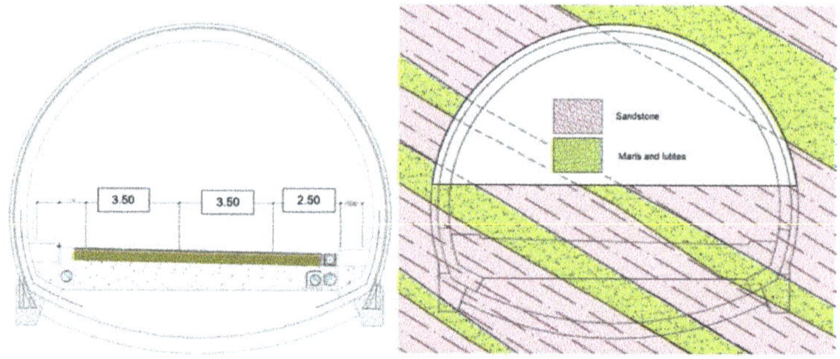

Figure 3. Cross section in aquifer influence zone (**left**). Alternation of sandstone levels with marls and lutites (**right**).

The construction of the project was divided into two stages: an initial project and a completion project, which was executed by the joint venture Viuda de Sainz and Lurpelan Tunnelling. In the completion project, 62 and 63 m were executed in top heading excavation (Axis 1 and Axis 2), as well as 160 and 164 m in bench excavation (Axis 1 and Axis 2). All the pending excavation was in the zone of influence of the aquifer (Figure 4).

2.2. Geological Study

The Urdinbide tunnel is geologically situated in the Basque–Cantabrian Basin, which is about 90 km wide and located between the Bay of Biscay and the Duero Depression. More specifically, the Urdinbide tunnel is located in the Biscay Synclinorium [24]. In the area, the synclinorium mostly forms a large structural valley of the same orientation, traversed by the Ibaizabal River, where the towns of Durango and Amorebieta are located, and which, when obliquely contacting the Cantabrian coast, gives rise to the Bilbao estuary. In Amorebieta, the north slope of the Ibaizabal River valley culminates in the Auztagane Range, crossed by the road to Gernika through the mountain pass of the same name. It is precisely this geographical feature that is intended to be overcome by the excavation and commissioning of the Urdinbide tunnel.

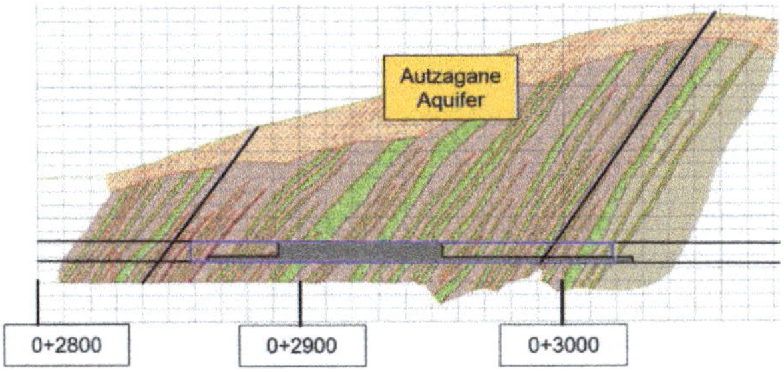

Figure 4. Location of the executed tunnel within the influence zone of the Autzagane aquifer. The figure shows the alternation of more or less permeable layers that explain its nature.

The geology of the Urdinbide massif is characterized by an alternation of sandstone and microconglomerate levels with marls and lutites (Figure 5). This alternating character, typical of flyschoid facies, is identified at all observation scales. The weathering of the massif affects both the sandstone and the lutitic and marl materials. In the first case, the dissolution of the carbonate cement in the sandstones results in sandy materials with the consequent increase in porosity and, therefore, a greater "penetrating" potential of weathering. In the case of lutitic and marl materials, the result of the alteration is clayey materials that do not substantially modify the low permeability of these materials and, therefore, have a lower capacity for alteration in depth.

Figure 5. Calcareous sandstone facies with isolated intercalations of marl material, recognized in the advance of Axis 1.

Vertically, weathering affects the sandstone and sand beds more deeply. Therefore, at the same horizontal level, there are beds or sections of more disintegrated and water-charged materials (the sandstone and sand beds) that vary alternately with other materials that are better geotechnically and more impermeable (the marl material), as shown in Figure 5. These geological characteristics determine the existence of a multilayered, basically confined aquifer. These characteristics are common to the aquifers found in the area.

2.3. Hydrogeological Study

During the project stage, a detailed hydrogeological characterization of the massif was carried out by the company EPTISA, determining the conceptual model of the aquifer's functioning, quantifying its water uses, and establishing the relationship between groundwater and the planned works before their execution. The hydrogeological study and its monitoring during the works were identified as critical elements in ensuring the final success of the project [25].

In order to achieve these objectives and establish the effects that tunnel drilling could have on the natural functioning of groundwater and its uses, various networks of piezometric and forometric control were installed.

Finally, the project included the creation of a series of numerical models of groundwater flow that have allowed validating the conceptual model of natural groundwater functioning and thus predict the effects of the tunnel on the aquifer's functioning and its water uses.

In order to carry out the hydrogeological study, existing information was analyzed first, including the construction projects that the promoter had drawn up in the work area, as well as general geological and hydrogeological studies that encompass the study area [26]. A total of 16 boreholes were drilled in which the piezometric level could be measured, with drilling depths ranging from 17 to 91 m.

In order to obtain the hydraulic parameters of the aquifer, several pumping tests were conducted. In this test, water is extracted in a controlled manner for a significant period of time, which is one of the most widely used methods for obtaining the aforementioned data [27]. In the research campaign carried out for the study, once the piezometric wells and pumping wells were drilled, three pumping tests were carried out in the wells called B-1, B-2, and B-3. Each test consisted of extracting water from the well and observing the variations in the water level depth produced in the pumping well itself and in all the existing perforations in the surrounding area that have been conditioned as piezometers. In each well, a stepped pumping stage, a constant flow stage, and a recovery stage were carried out (Figure 6). Mean permeability values of 3.5×10^{-6} m/s and a storage coefficient of 1.2×10^{-3} were obtained.

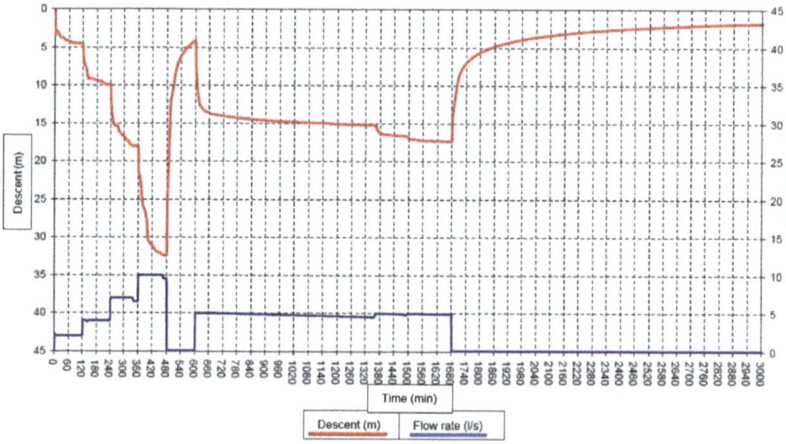

Figure 6. Graphical representation of pumping test data in well B-3.

During the hydrological study, 41 water points were identified. The main use of groundwater resources in the area closest to the south portal of the Urdinbide tunnel occurred in the headwaters of the Ategorri stream. In this sector, there were a total of 4 groundwater intakes, 3 surface water intakes, 1 distribution chamber, and a reservoir, connected by a complex network of pipes (Figure 7), all belonging to the water supply system of the Amorebieta-Etxano municipality.

A forometric network was also installed in order to allow continuous monitoring of the flow rates of the springs and intakes in the study area. To implement this control, the following infrastructure was established to determine the circulating flow rates: 3 electromagnetic flow meters with a diameter of 50 millimeters (two of them in the pipes entering the distribution chamber and the third in the pipe that goes from the Ategorri-A weir to the reservoir) and a 3-inch Parshall flume in the Ategorri stream upstream of the weir.

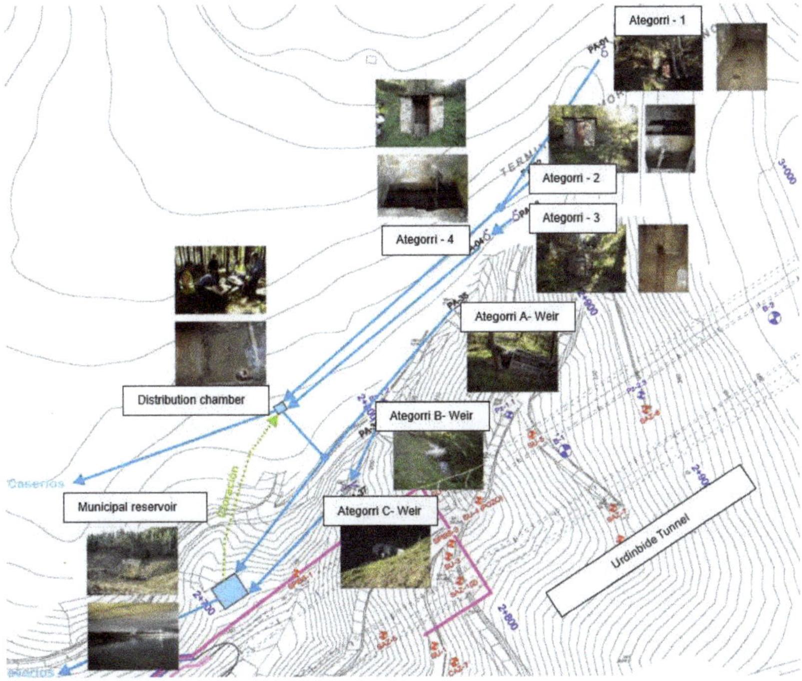

Figure 7. Inventory of main water points close to the south portal of the Urdinbide tunnel (4 groundwater intakes, 3 surface water intakes, 1 distribution chamber, and 1 reservoir).

2.3.1. Conceptual Hydrogeological Functioning Model

The hydrogeological conceptual model of the Autzagane aquifer was intended to convey the essence of the fundamental principles and basic functionality of the system, which it represents, and was defined by its nature, recharge and discharge mechanisms, and the groundwater flow scheme within it. The aquifer was divided into several hydrogeological sectors based mainly on the identified zones of groundwater discharge during the inventory of water points, although hydraulic continuity could exist between adjacent sectors. The hydrogeological sector of the Autzagane aquifer that is most directly related to the Urdinbide tunnel is the Ategorri sector.

The recharge of the Autzagane aquifer occurs through the infiltration of precipitation. The development of a significant layer of surface alteration and the moderately sloping terrain in the area promote a high infiltration coefficient in relation to the total usable rainfall (precipitation minus evapotranspiration). The average precipitation in the area is 1550 mm/year.

Regarding the discharge, the Autzagane aquifer is crossed by the heads of several streams that run in a NNE–SSW direction. These small valleys locally represent the axes of the lower relative elevation of the portion of the aquifer situated on both sides of them, and, therefore, constitute the discharge zones, through springs or diffuse discharges to the streams, of these sectors. The presence of lutitic and marly layers that separate the permeable sandstone layers of the aquifer causes the discharge of each sector to occur not at a single point corresponding to the intersection of the aquifer roof with the stream (as would be the case in a homogeneous aquifer) but at a series of springs along the stream related to each permeable level separated by less permeable intercalations.

This phenomenon was evident in the Ategorri sector, where there were four stepped springs in its headwaters (Ategorri-1 to 4). According to the data provided by the study's forometric monitoring network, the subterranean discharge in the Ategorri stream system would be around 4 L/s, of which 1.4 L/s were used for the partial supply of the municipality of Amorebieta-Etxano through the capture of the headwater springs of the Ategorri stream.

The circulation of groundwater from the recharge zones of the aquifer (the interfluves) to the discharge lines (the streams) occurred mainly through the permeable sandstone levels in the direction of the layers. Most of the groundwater flow occurred under confined conditions. The groundwater in the massif had a medium mineralization, and its hydrochemical facies was calcium bicarbonate [1].

2.3.2. Numerical model

The hydrogeological study included a numerical modeling of the behavior of groundwater in the massif crossed by the Urdinbide tunnel, which aimed to predict its response to the construction of the work. The Visual MODFLOW 6 software [28] was used to carry out the numerical model of the flow in the study area. This software was used because it is based on finite differences and can represent well the physical processes related to groundwater flow. The geometric scope of the modeled area can be seen in Figure 8. The aquifer was divided into discrete elements of a maximum width of 25 meters, achieving a mesh in which each cell represents a prism whose hydrogeological characteristics are constant throughout its volume.

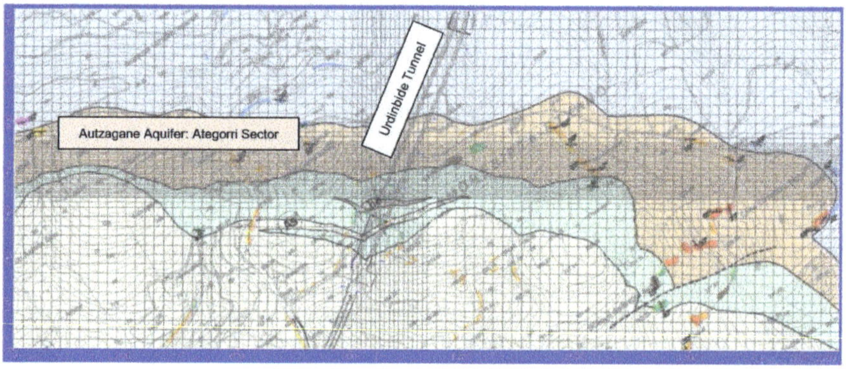

Figure 8. Model scope and surface discretization of the model in cells. Mesh with a maximum width of 25 m.

The spatial discretization of the model is shown in Figure 9.

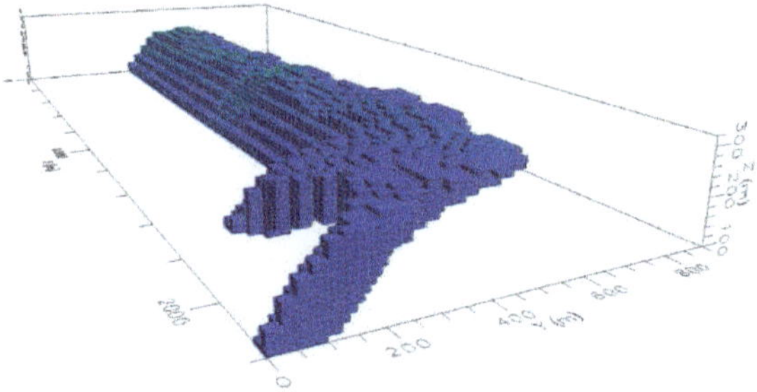

Figure 9. Three-dimensional view of the model showing the discretization in layers, using the Visual MODFLOW software.

The simplification of using a single value for each of the hydrogeological parameters corresponding to the set of geological materials that make up the modeled aquifer has been considered. By carrying out pumping tests in wells B2 and B3, representative values for the entire Autzagane aquifer were established, but during the model calibration phase, these values were slightly adjusted as it was considered that these corrections had a favorable impact on the model's fit to the known real functioning. Thus, the final adjusted hydraulic parameters are shown in Table 1.

Table 1. Adjusted hydraulic parameters.

Hydrostratigraphic Unit	Horizontal Permeability (m/s)	Vertical Permeability (m/s)	Storage Coefficient	Efficient Porosity (%)
Autzagane Aquifer	2.5×10^{-6}	2.5×10^{-6}	1×10^{-5}	5

Once the scope of the modeling had been established, the following actions on the system were considered: precipitation infiltration, discharges at the head of the Ategorri stream, discharges from other springs, and drainage from the existing tunnel of the Amorebieta-Etxano railway line. The model was adjusted in steady state, with the following water balance, which was consistent with the measured discharge in the entire Ategorri stream (Table 2).

Table 2. Annual water balance of the Autzagane aquifer in the natural regime after recalibration.

	Balance Items	Flow (L/s)
INPUT	Precipitation infiltration	9.1
OUTPUT	Ategorri stream (springs and diffuse discharges)	4.1
	Etxanosolo stream (springs and diffuse discharges)	1.5
	Torreburu spring	1.9
	Asketa spring	0.8
	Sastratxu spring	0.1
	Existing railway line tunnel	0.7
	Total	9.1

The calibrated model was used to establish the long-term response between two opposite scenarios: the scenario before the tunneling works and the limit scenario in a completely drained hypothesis. Figure 10 shows the isopiezometric lines of the model for the situation prior to the impact of the works in a section perpendicular to the Ategorri stream.

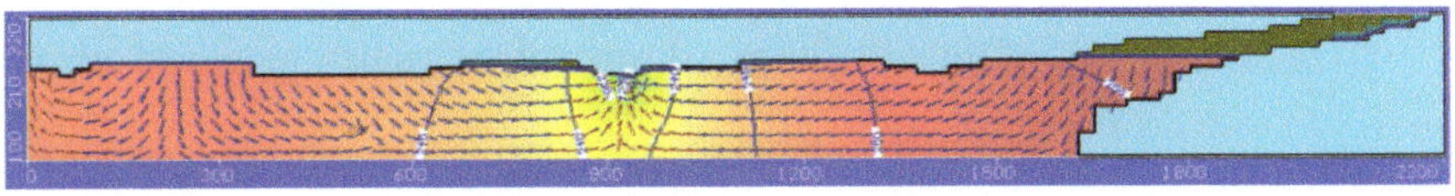

Figure 10. Isopiezometric lines for the situation before tunnel construction, using the Visual MODFLOW software.

Table 3 illustrates a comparison of flows between the natural situation (shown in Table 2) and the hypothesis of total draining tunnel. It should be noted that under this working hypothesis, the tunnel would be working as a scour outlet, with a flow rate of 6.6 L/s, leaving the Ategorri stream and its associated springs without flow.

Figure 11 shows the isopiezometric lines of the model for the situation considering a draining tunnel.

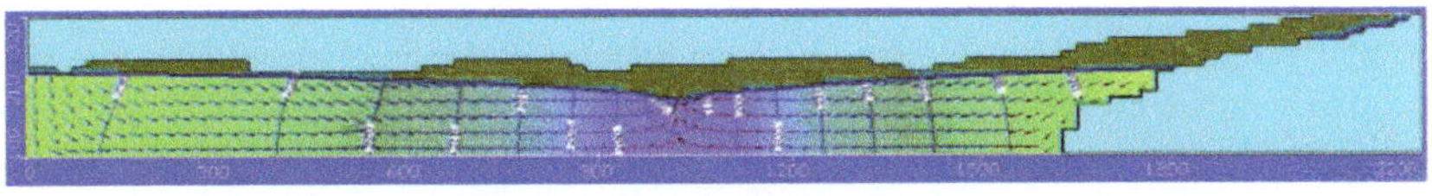

Figure 11. Isopiezometric lines for the hypothesis of draining tunnel, using the Visual MODFLOW software.

Table 3. Comparison between the natural regime and a draining tunnel regime.

	Balance Items	Flow in Natural Regime (L/s)	Flow with Draining Tunnel (L/s)	Variation
INPUT	Precipitation infiltration	9.1	9.1	-
OUTPUT	Ategorri stream (springs and diffuse discharges)	4.1	0	−100%
	Urdinbide tunnel	-	6.6	-
	Etxanosolo stream (springs and diffuse discharges)	1.5	0.8	−46%
	Torreburu spring	1.9	1	−47%
	Asketa spring	0.8	0.2	−75%
	Sastratxu spring	0.1	0	−100%
	Existing railway line tunnel	0.7	0.5	−28%
	Total	9.1	9.1	-

Finally, the hydrogeological model was used to obtain the limit flows that could be evacuated through the Urdinbide tunnels, guaranteeing the necessary flow in the Ategorri stream to supply the neighborhoods of Amorebieta-Etxano under normal conditions (1.4 L/s). The waterproofing solution for the tunnels, both during their construction and in service, had to be sufficient to guarantee a water discharge in the tunnel of less than these values (4.7 L/s).

2.3.3. Design of Waterproofing System

Due to the foreseeable impact of the tunnel construction on the Autzagane aquifer, the design project had planned the implementation of pre-excavation waterproofing injections from the advance fronts and the construction of a definitive watertight lining in the section that crosses the aquifer in order to guarantee its preservation. The main objectives of these injections were reducing the permeability of the rock mass to a level that the water infiltration into the tunnel would not exceed the maximum flow that could be extracted without significantly affecting the aquifer and the water catchment system and also achieving a crown of improved soil with a minimum thickness of 3 meters around the perimeter of the excavation (an increase in the GSI value by 20 points was required [29]).

The injections were executed from inside the tunnel. They consisted of truncated cone-shaped enclosures outside the excavation perimeter (Figure 12). The geometry of the injections was adapted to the orientation of the stratification surfaces of the rock mass so that the base of the truncated cone surface (plane defined by the end points of all the boreholes) was always a virtual plane parallel to this stratification. The injections were planned in two phases, top heading excavation and bench excavation, due to geometrical requirements. The treatment crowns were carried out in 4 stages, consisting of 30 drill holes in the top heading excavation phase and 19 drill holes in the bench excavation phase [30].

Stages 1 and 2, the outermost ones, were the first to obtain the outer wrapping. The other successive stages were implemented to intensify the degree of treatment. This system of execution allowed selective injection by sections and was used with satisfactory results in the tunnel sections already executed. The type of mixtures using cement and microcement was adjusted according to the gauged flow rates in each borehole, as well as the presence or absence of solid drags [31–33].

2.4. Hydrogeological Monitoring of the Tunnel during the Construction Phase

During the construction of the tunnel, an intensive hydrogeological monitoring was carried out to verify the working hypotheses. This monitoring consisted of continuous reading of the piezometric and forometric network installed in the study area. The construction of the tunnel caused a drop in water levels in the northern and central areas of the aquifer, and had a one-off impact on the municipal water intakes from the springs at the headwaters of the Ategorri stream.

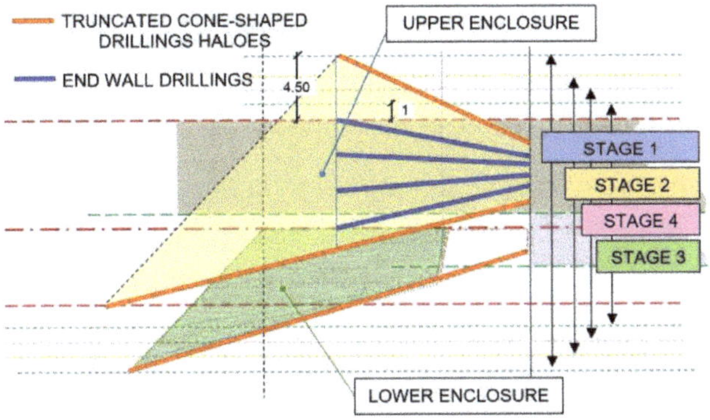

Figure 12. Schematic longitudinal profile of injection treatment. Stage situation.

The magnitude of these impacts was related to the phase of greatest activity during excavation, where potential temporary impacts during construction would be maximum.

The execution of preinjections and the advance excavation of tunnel tubes caused a depression of between 15 and 25 m compared with the over 50 m of maximum water column that existed above the tunnel level before the works (Figure 13).

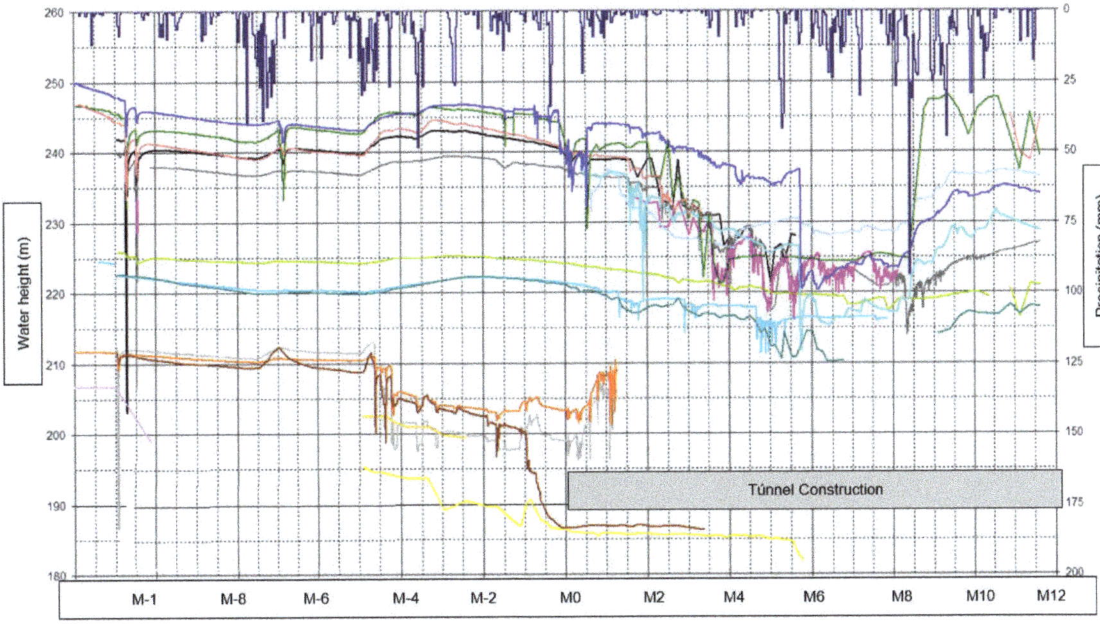

Figure 13. Evolution of the water level in the piezometers of the Autzagane aquifer. Evolution of the total precipitation in the area. The different colors correspond to different piezometers installed in the study area.

An extraordinarily dry and prolonged drought period that occurred in this area sensibly coinciding with the beginning of tunnel construction in the aquifer also contributed to the magnitude of these impacts.

The effectiveness of the waterproofing injections carried out, together with the recovery of the usual precipitation regime, partially reversed the impacts that the tunnel construction had caused on the natural functioning and use of the Autzagane aquifer. Since the commissioning of the tunnel in 2018, hydrogeological monitoring has been carried out during the operation phase. Since then, the maximum flows measured through the tunnel have been less than 2 L/s, far from the limit imposed to guarantee the functioning of the aquifer and the municipal water catchment (Table 4).

Table 4. Determination of the limit flow values ensuring the water supply in the Ategorri stream.

	Balance Items	Flow in Natural Regime (L/s)	Flow Limit Values (L/s)	Variation
INPUT	Precipitation infiltration	9.1	9.1	-
OUTPUT	Ategorri stream (springs and diffuse discharges)	4.1	1.4	−65%
	Urdinbide tunnel	-	4.7	-
	Etxanosolo stream (springs and diffuse discharges)	1.5	0.8	−46%
	Torreburu spring	1.9	1.3	−31%
	Asketa spring	0.8	0.3	−63%
	Sastratxu spring	0.1	0	−100%
	Existing railway line tunnel	0.7	0.6	−14%
	Total	9.1	9.1	-

3. Results

From the monitoring of the hydrogeological performance of the aquifer during the construction of the tunnel, it can be established that if precautions had not been taken to carry out prior waterproofing injections before tunnel excavation in the Autzagane aquifer, the water level in it would have dropped to the elevation of the tunnel, all the springs and other underground discharges towards the Ategorri stream would have dried up, and the excavation itself would have drained practically all the resources of this sector of the aquifer.

The effectiveness of sequential waterproofing prior to excavation was high, as it has allowed the recovery of piezometric levels and the resurfacing of affected springs as soon as precipitation regimes returned to normal once the tunnel construction was finished. The injection system maintained the original compartmentalization of the aquifer, as demonstrated by the registered piezometry, without the tunnel unifying the piezometric levels of the different subsectors.

In general terms, the natural hydrogeological functioning of the aquifer has been preserved. The waterproofing system used preserved not only the aquifer but also the original drainage points that constitute the springs of the Ategorri stream.

The waterproofing solution, consisting of a combination of cement and microcement injections specifically designed for the characteristics of the study area, has therefore made it possible to comply with the main requirement imposed by the project's environmental authorization, no significant impact on the Autzagane aquifer.

Author Contributions: Conceptualization, J.-M.B., I.E. and M.G.; methodology, J.-M.B., I.E. and M.G.; formal analysis, J.-M.B. and I.E.; writing—original draft preparation, J.-M.B.; writing—review and editing, J.-M.B., I.E. and M.G.; supervision, J.-M.B. All authors have read and agreed to the published version of the manuscript.

Funding: This research received no external funding.

Institutional Review Board Statement: Not applicable.

Informed Consent Statement: Not applicable.

Data Availability Statement: The data presented in this study are available on request from the corresponding author. The data are not publicly available due to confidentiality.

Acknowledgments: The authors want to acknowledge the Interbiak company for the facilities provided for the development of the research.

Conflicts of Interest: The authors declare no conflict of interest.

References

1. Interbiak. *Project for Completion of the Works in the Amorebieta-Muxika Section*; Diputación Foral de Bizkaia, Interbiak: Bilbao, Spain, 2015.
2. Ilyushin, Y.V.; Asadulagi, M.-A.M. Development of a Distributed Control System for the Hydrodynamic Processes of Aquifers, Taking into Account Stochastic Disturbing Factors. *Water* **2023**, *15*, 770. [CrossRef]
3. Martirosyan, A.V.; Kukharova, T.V.; Fedorov, M.S. Research of the Hydrogeological Objects' Connection Peculiarities. In Proceedings of the 2021 IV International Conference on Control in Technical Systems (CTS), Saint Petersburg, Russia, 21–23 September 2021; pp. 34–38. [CrossRef]
4. Martirosyan, A.V.; Martirosyan, K.V.; Mir-Amal, A.M.; Chernyshev, A.B. Assessment of a Hydrogeological Object's Distributed Control System Stability. In Proceedings of the 2022 Conference of Russian Young Researchers in Electrical and Electronic Engineering (ElConRus), Saint Petersburg, Russia, 25–28 January 2022; pp. 768–771. [CrossRef]
5. Wang, S.; Gao, Z.; Wang, Z.; Wu, X.; An, Y.; Ren, X.; He, M.; Wang, W.; Liu, J. Hydrodynamic characteristics of groundwater aquifer system under recharge and discharge conditions. *Arab. J. Geosci.* **2020**, *13*, 859. [CrossRef]
6. Lasagna, M.; Mancini, S.; De Luca, D. Groundwater hydrodynamic behaviours based on water table levels to identify natural and anthropic controlling factors in the Piedmont Plain (Italy). *Sci. Total Environ.* **2020**, *716*, 137051. [CrossRef] [PubMed]
7. Hu, L.Y.; Chugunova, T. Multiple-point geostatistics for modeling subsurface heterogeneity: A comprehensive review. *Water Resour. Res.* **2008**, 44. [CrossRef]
8. Tziritis, E.; Sachsamanoglou, E.; Aschonitis, V. Assessing Groundwater Evolution with a Combined Approach of Hydrogeochemical Modelling and Data Analysis: Application to the Rhodope Coastal Aquifer (NE Greece). *Water* **2023**, *15*, 230. [CrossRef]
9. Dawson, C. A continuous/discontinuous Galerkin framework for modeling coupled subsurface and surface water flow. *Comput. Geosci.* **2008**, *12*, 451–472. [CrossRef]
10. Harter, T. Finite-size scaling analysis of percolation in three-dimensional correlated binary Markov chain random fields. *Phys. Rev. E* **2005**, *72*, 026120. [CrossRef]
11. Zammouri, M.; Brini, N. Efficiency of Artificial Groundwater Recharge, Quantification Through Conceptual Modelling. *Water Resour. Manag.* **2020**, *34*, 3345–3361. [CrossRef]
12. Condon, L.E.; Kollet, S.; Bierkens, M.F.P.; Fogg, G.E.; Maxwell, R.M.; Hill, M.C.; Fransen, H.H.; Verhoef, A.; Van Loon, A.F.; Sulis, M.; et al. Global groundwater modeling and monitoring: Opportunities and challenges. *Water Resour. Res.* **2021**, *57*, e2020WR029500. [CrossRef]
13. Condon, L.E.; Hering, A.S.; Maxwell, R.M. Quantitative assessment of groundwater controls across major US river basins using a multi-model regression algorithm. *Adv. Water Resour.* **2015**, *82*, 106–123. [CrossRef]
14. Vázquez-Suñé, E.; Capino, B.; Abarca, E.; Carrera, J. Estimation of Recharge from Floods in Disconnected Stream Aquifer Systems. *Groundwater* **2007**, *45*, 579–589. [CrossRef] [PubMed]
15. Kolymbas, D.; Wagner, P. Groundwater ingress to tunnels—The exact analytical solution. *Tunn. Undergr. Space Technol.* **2007**, *22*, 23–27. [CrossRef]
16. Shin, J.H.; Addenbrooke, T.I. A numerical study of the effect of groundwater movement on long-term tunnel behaviour. *Geotechnique* **2002**, *52*, 391–403. [CrossRef]
17. Nazarchuk, A. Water Intrusion in Underground Structures. Master's Thesis, Massachusetts Institute of Technology, MCambridge, MA, USA, 12 June 2008.
18. Dvanajščak, D.; Ratej, J.; Jovičić, V. Sustainability of Water Resources in Karst Undermined by Tunneling: A Case Example. *Sustainability* **2022**, *14*, 732. [CrossRef]
19. Saito, H.; Date, K.; Narita, N.; Yamamoto, T.; Yokota, Y.; Koizumi, Y. Pre-grouting and tunnel excavation of watertight structure section. In Proceedings of the ISRM International Symposium—8th Asian Rock Mechanics Symposium, Sapporo, Japan, 14–16 October 2014; pp. 1184–1189.
20. Golian, M.; Abolghasemi, M.; Hosseini, A.; Abbasi, M. Restoring groundwater levels after tunneling: A numerical simulation approach to tunnel sealing decision-making. *Hydrogeol. J.* **2021**, *29*, 1611–1628. [CrossRef]
21. Trinh, N.Q. Controlling of groundwater inflow: From Norwegian sub-sea tunnels to tunnel in large cities. In Proceedings of the 13th World Conference of ACUUS: Advances in Underground Space Development, Singapore, Singapore, 7–9 November 2012; pp. 1170–1180.
22. Rodríguez-Loinaz, G.; Amezaga, I.; Onaindia, M. Efficacy of Management Policies on Protection and Recovery of Natural Ecosystems in the Urdaibai Biosphere Reserve. *Nat. Areas J.* **2011**, *31*, 358–367. [CrossRef]
23. Barton, N. Ground Stabilisation. The why's and how's of high pressure grouting—Part 1. *Tunn. Tunn. Int.* **2014**, 28–30.
24. Meschede, M. The tectonic and sedimentary development of the Biscay synclinorium in Northern Spain. *Geol. Rundsch.* **1987**, *76*, 567–577. [CrossRef]

25. Lo Russo, S.; Taddia, G.; Cerino, E. Tunnelling and groundwater interaction: The role of the hydrogeological monitoring. *Geoing. Ambient. E Min.* **2015**, *146*, 37–44.
26. EVE. *Hydrogeological Map of the Basque Country at a Scale of 1:100.000*; EVE: Bilbao, Spain, 1996.
27. Manoj, P.; Madan, K. Estimation of Aquifer Parameters from Pumping Test Data by Genetic Algorithm Optimization Technique. *J. Irrig. Drain.* **2003**, *129*, 348–359. [CrossRef]
28. Visual Modflow Flex 9.0. Available online: https://www.waterloohydrogeologic.com/products/visual-modflow-flex/ (accessed on 30 April 2023).
29. Marinos, V.; Marinos, P.; Hoek, E. The geological strength index: Applications and limitations. *Bull. Eng. Geol. Environ.* **2005**, *64*, 55–65. [CrossRef]
30. Baraibar, J.M.; Gil, M.; Escobal, I. Design of microcement-based injections in highly porous media. Urdinbide Tunnel. In Proceedings of the ITA-AITES World Tunnel Congress, Copenhagen, Denmark, 2–8 September 2022.
31. Haugsand, M. Hydraulic Jacking and Pressure Distribution During Rock Mass Grouting with Cement Based Grouts. In Proceedings of the ISRM 9th Nordic Grouting Symposium, Helsinki, Finland, 2–3 September 2019.
32. Escobal Marcos, I.; Alvarez-Fernandez, M.I.; Prendes-Gero, M.B.; Gonzalez-Nicieza, C. Designing Cement-Based Grouting in a Rock Mass for Underground Impermeabilization. *Energies* **2021**, *14*, 4062. [CrossRef]
33. González-García, J.; González-Nicieza, C.; Álvarez-Fernández, M.-I.; Prendes-Gero, M.-B. Injection Treatment for Tunneling Excavation in Sandy Soils with High Fines Content. *Energies* **2021**, *14*, 6930. [CrossRef]

applied sciences

MDPI

Article

A Numerical Simulation of the Subsidence Reduction Effect of Different Grouting Schemes in Multi-Coal Seam Goafs

Hai Wang [1,2], Yan Qin [1,*], Yuxi Guo [1] and Nengxiong Xu [1]

[1] School of Engineering and Technology, China University of Geosciences (Beijing), Xueyuan Road 29, Beijing 100083, China; wh_rocks@163.com (H.W.); 2102200103@cugb.edu.cn (Y.G.); xunengxiong@cugb.edu.cn (N.X.)

[2] Shanxi Traff Layout Reconnaissance Design Inst., Co., Wuluo St. 27, Taiyuan 030032, China

* Correspondence: qinyancugb@cugb.edu.cn; Tel.: +86-(010)-82322627

Abstract: Grouting is the most widely used technology for treating coal goafs. In this study, a numerical simulation method was used to establish a model of multi-seam goafs with different spacing conditions to investigate the subsidence reduction effects of various grouting schemes on multi-coal seam goafs. By varying the range and opportunity of grouting treatments, the effects of coal seam spacing, grouting treatment range, and grouting opportunity on subsidence reduction were analyzed. The results showed that: (1) With constant overburden (OB), the subsidence reduction ratio of the subgrade center increases linearly as the interburden (IB) decreases ($1 \leq OB/IB \leq 2$), then increases exponentially ($2 < OB/IB$), and eventually becomes stable. (2) When treatment is conducted based on the half-width of the subgrade, the width of the subgrade, and the range of the trapezoid, residual surface subsidence tends to adopt an inclined 'W'-shape in open cutting. The surface residual subsidence exhibits a symmetrical 'W'-shape when full-range grouting is adopted. (3) For a multi-coal seam goaf with longer mining stoppage time, the subsidence reduction ratio of the subgrade center is lower, and it is exponentially related to the grouting opportunity. As the grouting opportunity is extended and OB/IB decreases, the subsidence reduction ratio of the subgrade center declines exponentially.

Keywords: multi-coal seam goaf; range of grouting; opportunity of grouting; subsidence reduction effect; numerical simulation

Citation: Wang, H.; Qin, Y.; Guo, Y.; Xu, N. A Numerical Simulation of the Subsidence Reduction Effect of Different Grouting Schemes in Multi-Coal Seam Goafs. *Appl. Sci.* **2023**, *13*, 5522. https://doi.org/10.3390/app13095522

Academic Editor: Bing Bai

Received: 9 April 2023
Revised: 26 April 2023
Accepted: 27 April 2023
Published: 28 April 2023

1. Introduction

The subsidence area of coal mining in China has exceeded 8000 km^2 and is still growing at a rate of 200 km^2 per year [1–3]. Full-pressure grouting filling, which is an important engineering method, can effectively strengthen the broken rock structure of goafs and reduce the occurrence of deformation disasters in goafs [4–11], and it is widely used in goaf treatment engineering [12–20]. Many experts and scholars have studied the grouting treatment of goafs from the aspects of slurry diffusion, slurry stone strength, and grouting scheme design and have achieved rich research results.

The grouting slurry flows and diffuses under pressure, filling the gaps between broken rock masses. Wang et al. [1] revised a theoretical formula for slurry diffusion based on the distribution characteristics of fractures in a goaf and fracture zones of a single coal seam, as well as the effect of the superposition of porous grouting and the viscosity and diffusion path of the slurry. Yuan et al. [21] independently designed a visual simulation system for grouting in a single coal seam goaf and used it to study the pressure distribution, diffusion radius, and thickness of the stone body of the slurry in the rock layer. The slurry flow showed an approximately elliptical diffusion range under pressure, and the degree of grouting reinforcement increased with distance from the injection borehole, providing a basis for the optimization of the design of grouting in a single coal seam goaf. Yu et al. [22], who took the engineering of a single coal seam goaf roof in weakly cemented siltstone, was taken as the research background. Methods of field investigation, laboratory testing, and theoretical analysis were adopted. Based on a detailed study of the diffusion of grouting materials, a combined support scheme based on grouting reinforcement and the use of anchor cables was proposed and applied in practical projects. When the slurry diffuses to the designated area for goaf treatment, grouting stones form in the cavities and in areas of the goaf under pressure. The strength of the grouting stones is an important indicator used to measure the effectiveness of grouting reinforcement. Liu et al. [23] conducted grouting tests on the fractured surface of the caving rock mass to solve engineering problems such as rock mass caving in a single coal seam goaf and studied the tensile characteristics of the goaf rock mass after grouting reinforcement. Xie et al. [24] applied numerical simulations to analyze the effective pre-stress field distribution of broken roof and grouting roof anchor cables in a multi-seam goaf. The strength of grouting stone bodies in a multi-coal seam goaf was monitored using experimental laboratory methods, and a treatment method of using grouting anchor cables to fix the weak surface of the roof plate and seal roof cracks was proposed. Han et al. [25] revealed the stress-hardening characteristics and load-bearing mechanism of grouting stones in uniaxial compression tests by conducting experiments on their physical properties. Modoni et al. [26] proposed a method for predicting the strength of grouted stone bodies based on their interaction with surrounding rock and soil. Zong et al. [27] conducted uniaxial compression tests on fractured rock masses after grouting and found the influences of grouting stone strength, deformation characteristics, and failure mode on the grouting reinforcement effect. They also found that the failure mode of the sample changed from brittle failure to plastic failure. The above research results provide a basis for grouting treatment technology, and experts and scholars have conducted research on the effect of grouting treatment plans on subsidence reduction. Wang et al. [28] combined numerical simulation and laboratory experiments and proposed a grouting method for preventing and controlling safety hazards in a single coal seam goaf by using pre-drilling, full-hole intubation, extended sealing, and multi-stage drilling for grouting. Li et al. [19] used numerical simulation software to simulate the reinforcement effects of two grouting methods—full-range grouting and strip grouting—in a single coal seam goaf. They found that a reasonable grouting plan can effectively reduce residual deformation of the goaf and its overlying rock mass, improve adverse stress conditions, and achieve the goal of effective grouting. Based on a theory of mining subsidence, Deng et al. [29] studied grouting filling technology to control the residual subsidence of a single coal seam goaf and concluded that under a certain mining depth, goaf grouting can be filled with banded grouting. The grouting hole should be arranged within 20 m of the edge of the working face, in the direction of the steeply inclined coal seam and the fault development area to fill the voids and under-compacted areas at the edge of the goaf. To explore the grouting scheme of multi-coal seam goafs, Chen et al. [30] studied the grouting treatment technology of multi-coal seam goafs by combining various geophysical exploration methods and drilling methods. According to the spacing of each coal seam goaf, the grouting horizon was divided, and the grouting technology and scheme of "pressure less gravity flow and pressurized diffusion" were used to reinforce the goaf. He et al. [31] used methods such as on-site investigation, laboratory testing, and numerical simulation to study the deformation and failure modes and stress distribution characteristics of multi-seam goafs at close range under double-thick coal seam mining conditions. A zoning grouting treatment method centered on "high pressure water jet, asymmetric high strength cable beam net, three hole anchor cable group and roof grouting" was proposed.

Most of the above studies focused on the treatment of single-coal seam grouting. Due to the large burial depth and multiple layers of goaf in multi-coal seam goafs, if grouting is carried out according to the recommended standard [32] and the grouting range is determined based on the rock movement angle, then it causes the grouting range to be too large, which affects the ecological protection and restoration of the grouting area and may not necessarily achieve the expected subsidence reduction effect. The selection of grouting treatment opportunity also directly affects the residual subsidence level of multi-seam goafs.

The goal of this study was to further clarify the control of the effects of different grouting schemes on the deformation of multi-coal seam goafs. A model of a multi-coal seam goaf with different spacing conditions was designed using a numerical simulation method. By changing the grouting range and opportunity, the long-term deformation of multi-coal seam goafs with different spacings after different grouting schemes was calculated, and the subsidence reduction effect of different grouting schemes on the deformation of multi-coal seam goafs was analyzed. The results can provide more scientific and reliable guidance for the treatment of multi-seam goafs.

2. Proposed Method

This study establishes a numerical model of multi-coal seam goafs with different spacing, selects an appropriate constitutive model, considers different grouting treatment ranges and opportunities in the multi-coal seam goaf grouting treatment plan, studies the grouting subsidence reduction effect of different grouting schemes on multi-coal seam goafs, and reveals the influences of grouting opportunity and range selection on the treatment of multi-coal seam goafs.

2.1. *Establishment of a Numerical Model of a Multi-Seam Goaf*

Previous studies have shown that FLAC calculation software can effectively simulate coal seam mining and grouting treatment problems [33,34]. The Moh–Coulomb model, a double-yield model, and Burgers' model can effectively simulate coal seam mining, grouting treatment, and long-term deformation of goaf. The numerical model of the multi-coal seam goaf was a pseudo-3D model with a thickness of 4 m for both the upper and lower coal seams, and the overlying rock lithology was medium-hard rock. According to the standard [33], the thickness of the caving zone was calculated to be 10 m. To ensure sufficient mining, the lengths of the upper and lower coal seam working faces were taken as 1000 m. In each working condition, the thickness of the overburden above the upper coal seam was taken as 100 m, and the thickness of the bottom plate below the lower coal seam was taken as 40 m, as shown in Figure 1.

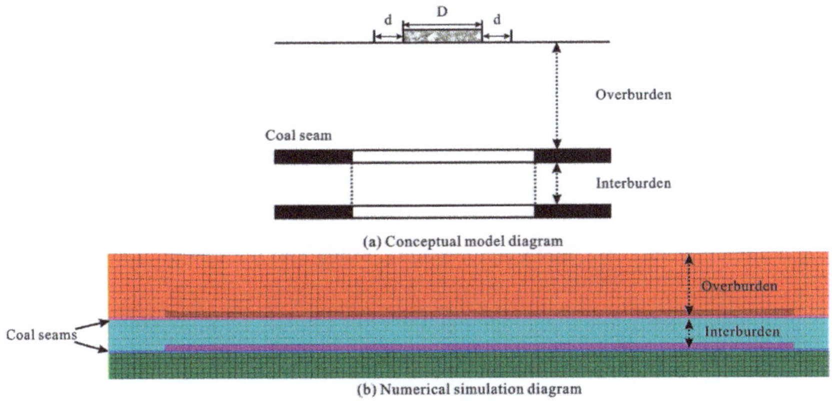

Figure 1. Completely overlapping mining schematic diagram.

Ten sets of multi-coal seam goaf spacing models were designed, all of which used the same overburden (OB) and different the interburden (IB). The dimensions of the upper and lower coal seam working faces were the same, but the burial depth in the longitudinal direction differed. The spacing information of the coal seams is shown in Table 1.

Table 1. Design table for spacing of multi-coal seam goaf.

Case Number	Overburden (OB)/m	Interburden (IB)/m	Working Face Length
1	100	10	1000 m for both upper and lower coal seam working faces
2		20	
3		30	
4		40	
5		50	
6		60	
7		70	
8		80	
9		90	
10		100	

This model adopted hexahedral mesh generation with a mesh size of 10 m. The displacement in the x-direction of the left and right boundaries of the model, the displacement in the y-direction of the front and rear boundaries, and the displacement in the z-direction of the lower boundary were fixed. The upper boundary was a free boundary. In the calculation, the self-weight of the rock mass was considered, the value of acceleration due to gravity in the model was 9.80 m/s^2, and the mining step distance was 10 m.

(1) Calculation models

For unexcavated coal seams and surrounding rocks, the Mohr–Coulomb model is used for calculation. During the equivalent mining process, a double-yield model is used to simulate the compaction process of collapsed rock masses in the caving zone. After the mining is completed, Burgers' model is used to dynamically assign creep parameters to the units within the corresponding interval range to calculate the creep process of the collapsed rock mass in the caving zone.

(2) Calculation parameters

To prevent the impact of different lithological combinations on the calculation results, the calculation parameters for the surrounding rock in this study were unified as medium-hard rock calculation parameters. Based on relevant laboratory experiments and engineering experience, the engineering analogy method was used to determine the calculation parameters for each layer in this study, as shown in Table 2. The equivalent mining method was used to assign the mined coal seam and the rock mass in the caving zone to a double-yield model to simulate the compaction process of the fractured rock mass. The calculation parameters are shown in Table 3 [33]. After mining in the goaf is stopped, the fractured rock mass and grouting stone mass undergo creep deformation under the pressure of the overlying rock layer. Burgers' model must be assigned to the designated area, and the calculation parameters are shown in Tables 4 and 5 [34].

Table 2. Calculation parameters for each layer.

Stratum	ρ (kg/m^3)	E_T (GPa)	ν	c (MPa)	φ (°)	σ_t (MPa)
Overlying rock	2750	3.20	0.25	1.00	30	0.35
Coal	1380	0.45	0.34	0.17	20	0.05
Floor	2750	3.20	0.25	1.00	30	0.35

Table 3. Calculation parameters of the double-yield model [33].

ρ (kg/m^3)	Maximum Bulk Modulus (GPa)	Maximum Shear Modulus (GPa)	c (MPa)	φ (°)	Plastic Modulus Multiplier
1680	2.24	1.41	0	15	3.0

Table 4. Calculations of creep parameters of medium-hard rock [34].

Creep Parameters	Compressive Stress Range/MPa	Parameter Calculation Formula
K/MPa	$0 < \sigma_{zz}$	$K = 6.35\ \sigma_{zz}{}^{1.78}$
G_m/MPa	$0 < \sigma_{zz} < 6.16$	$G_m = -6.82\ \sigma_{zz} + 344$
	$6.16 < \sigma_{zz} < 9.24$	$G_m = 2.92\ \sigma_{zz} + 284$
	$9.24 < \sigma_{zz}$	$G_m = -747\ \sigma_{zz} + 380$
G_k/MPa	$0 < \sigma_{zz}$	$G_k = 27.27\ \sigma_{zz} - 59$
η_m/MPa·h	$0 < \sigma_{zz} < 6.16$	$\eta_m = 11{,}613.96\ \sigma_{zz} + 11{,}446$
	$6.16 < \sigma_{zz} < 9.24$	$\eta_m = -5721.46\ \sigma_{zz} + 118{,}232$
	$9.24 < \sigma_{zz}$	$\eta_m = -850.812\ \sigma_{zz} + 73{,}227.5$
η_k/MPa·h	$0 < \sigma_{zz} < 6.16$	$\eta_k = -26.62\ \sigma_{zz} + 4728$
	$6.16 < \sigma_{zz} < 9.24$	$\eta_k = 55.19\ \sigma_{zz} + 4224$
	$9.24 < \sigma_{zz}$	$\eta_k = -42.86\ \sigma_{zz} + 4338$

Table 5. Calculations of creep parameters of grouting stone [34].

Creep Parameters	Parameter Calculation Formula
K/MPa	$K = 5.18697 \times e^{\frac{\sigma 1}{3.75492}} + 31.74$
G_m/MPa	$G_m = 40.4643 \times \sigma_1 + 41.5749$
G_k/MPa	$G_k = -1111.16204 + 776.47378 \times \sigma_1 - 81.25659 \times \sigma_1{}^2 + 2.97386 \times \sigma_1{}^3 \times 10^{-2}$
V_m/MPa·h	$V_m = 9 \times 10^9$
V_k/MPa·h	$V_k = 17088.0167 - 9300.51676 \times \sigma_1 + 1337.43539 \times \sigma_1{}^2$

2.2. Design of Different Grouting Ranges

Figure 2 shows a schematic diagram of the treatment range of the goaf in a horizontal multi-coal seam goaf. In this simulation plan, four types of grouting treatment ranges are designed:

(1) Treatment within the half-width of the subgrade. As shown in the blue line area in Figure 2, the treatment width of the goaf caving zone in both the upper and lower coal seams was 40 m.
(2) Treatment within the width of the subgrade. As shown in the green line area in Figure 2, the treatment width of the goaf caving zone in both the upper and lower coal seams was 80 m.
(3) The standard suggests the treatment of the trapezoidal range [32]. This treatment is as shown in the red line area in Figure 2 below.
(4) Full-range treatment of the working face. As shown in the black line area in Figure 2, grouting treatment was carried out throughout the goaf caving zone of the upper and lower coal seams, with a treatment width of 1000 m.

This numerical simulation studies the impact of different grouting treatment ranges on the control and effect of subsidence reduction in multi-seam goafs by changing the grouting width within the range of the caving zone in the multi-seam goaf. Taking the grouting treatment condition with a spacing of 10 m between the upper and lower coal seams as an example, the design schemes for different grouting treatment ranges are shown in Table 6.

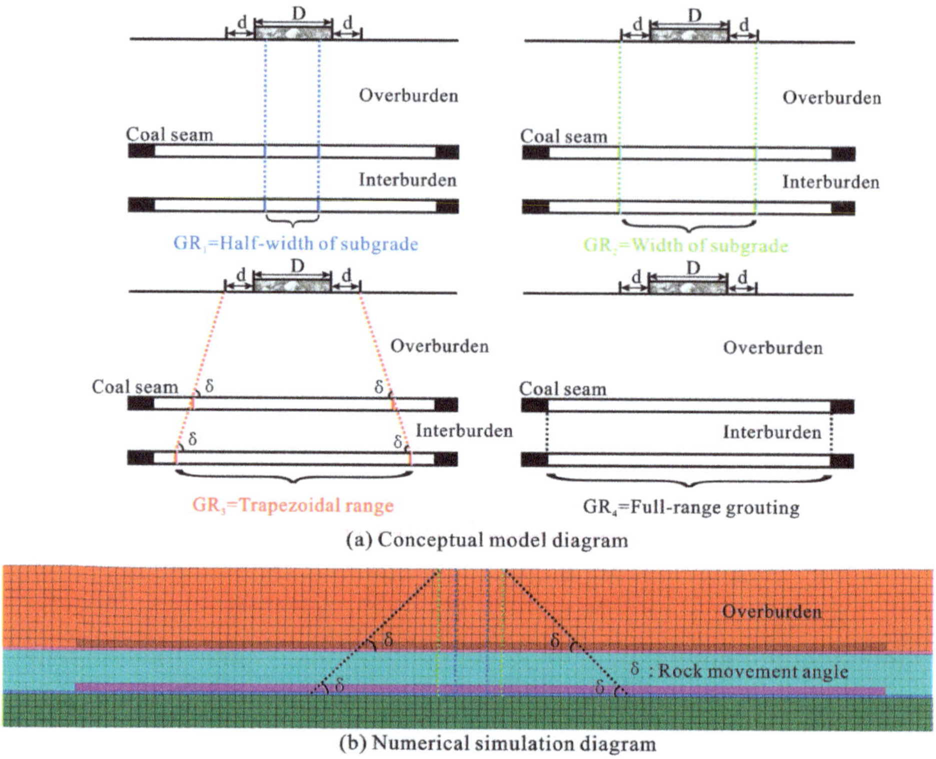

Figure 2. Diagram of the multi-seam goaf grouting ranges.

Table 6. Design of different grouting treatment ranges.

Interburden/m	Grouting Opportunity	Grouting Ranges (GR)
IB = 10	Immediate grouting	GR_1 = Half-width of subgrade GR_2 = Width of subgrade GR_3 = Trapezoidal range GR_4 = Full-range grouting

2.3. Design of Different Grouting Opportunity Schemes

This study uses numerical simulation to study the effect of grouting opportunity on the subsidence reduction effect of multi-seam goaf grouting. Taking the full-range grouting treatment condition with a spacing of 50 m between the upper and lower coal seams as an example, the design of different grouting opportunity schemes is shown in Table 7.

Table 7. Design of different grouting opportunities.

Interburden/m	Grouting Ranges	Grouting Opportunity (Time Interval between Stopping Mining)/Year
IB = 50	GR_4	GO_1 = 0.0 GO_2 = 0.5 GO_3 = 1.0 GO_4 = 1.0 GO_5 = 2.0 GO_6 = 2.5 GO_7 = 3.0

First, the upper coal seam working face is excavated, and then Burgers' model is applied to the rock mass in the caving zone to calculate the long-term deformation within one year. Then, the working face of the lower coal seam is excavated, and different grouting opportunities are selected to carry out comprehensive grouting treatment on the goaf of the upper and lower coal seams. Finally, the long-term deformation of the surface after 10 years is calculated, and the subsidence pattern of the subgrade center is recorded. From this, it is possible to simulate the subsidence of the subgrade center and the amount of grouting reduction when grouting is carried out in the goaf at different points in time after the completion of multi-coal seam mining and to obtain the impact of grouting opportunity on the grouting reduction effect in the goaf of multi-coal seams.

3. Results and Discussion

3.1. Results of Calculations of Grouting Subsidence Reduction in Goaf Areas with Different Spacing

Based on the method described earlier, the residual subsidence of the surface after grouting was simulated under different coal seam spacing conditions. The calculation results are shown in Figure 3, which are the residual subsidence cloud maps of grouting with coal seam spacings of 10 m, 30 m, 50 m, 70 m, and 90 m. From the residual subsidence cloud maps of grouting, it can be concluded that under different coal seam spacing conditions, after grouting treatment, the residual subsidence value of the surface near the stop mining line and the opening hole is relatively large. In contrast, the residual subsidence value of the surface near the center of the roadbed is relatively small, and as the burial depth of goaf in multi-coal seam increases, the residual subsidence after surface grouting treatment gradually increases. When the distance between coal seams reaches 70 m, there is little difference in the residual subsidence cloud maps of multi-coal seam goaf, and the influence of coal seam distance on the grouting treatment effect of multi-coal seam goaf is weakened. The cantilever beam structure of the roof near the opening and stopping lines in the goaf of the multi-coal seam is relatively long, resulting in under-compaction in this area. Therefore, there will still be some subsidence after grouting treatment compared to the compacted area below the surface subsidence basin. The greater the burial depth of the goaf in the multi-coal seam, the greater the pressure on the goaf from the overburden, resulting in more significant residual subsidence after treatment [22].

According to the calculation results in Figure 3, the subsidence of the subgrade center under grouted and non-grouted conditions for the spacing conditions of each coal seam are calculated and recorded as L_1 and L_2, respectively. Then, the grouting subsidence reduction for different coal seam spacing conditions is calculated according to Equation (1) for ΔL. According to Equation (2), the grouting subsidence reduction rate R is obtained for different coal seam spacing conditions.

$$\Delta L = L_1 - L_2 \tag{1}$$

$$R = \Delta L / L_1 \tag{2}$$

The relationship between the center subsidence reduction ratio R of the subgrade and the distance between coal seams is shown in Figure 4. From the figure, it can be concluded that when the spacing between coal seams is 10 m, the grouting has the best effect on reducing subsidence, and the ratio of subsidence reduction at the center of the subgrade can reach 81.1%. As the spacing between coal seams continues to increase, the ratio of subsidence reduction at the center of the subgrade gradually decreases after grouting treatment. When the spacing between coal seams is greater than 80 m, the rate of decrease in the center subsidence rate of the subgrade gradually decreases with the increase in the spacing between coal seams, and the curve tends to be flat. During the process of increasing the spacing between coal seams from 10 m to 80 m, the reduction ratio significantly decreases from 81.1% to 50.1%. Afterwards, the reduction ratio at the center of the subgrade is approximately 46%, and there is no significant change with the increase in the spacing between coal seams.

Residual subsidence time : 10 years ;

Selection of grouting scheme:

①Coal seam spacing selection scheme IB=10、30、50、70、90m ;

②Grouting treatment range selection scheme GR_4 ;

③Grouting opportunity selection scheme GO=0year ;

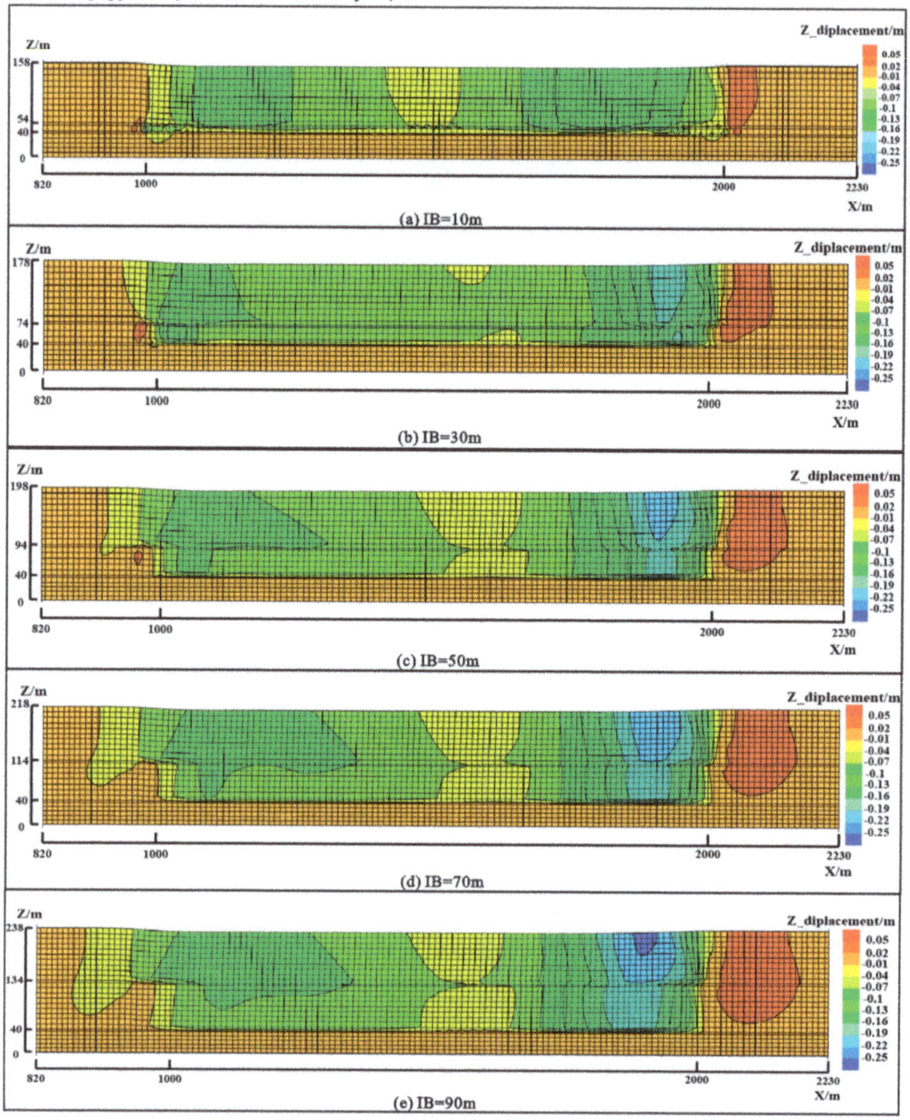

Figure 3. Grouting residual subsidence nephogram of different coal seam spacings.

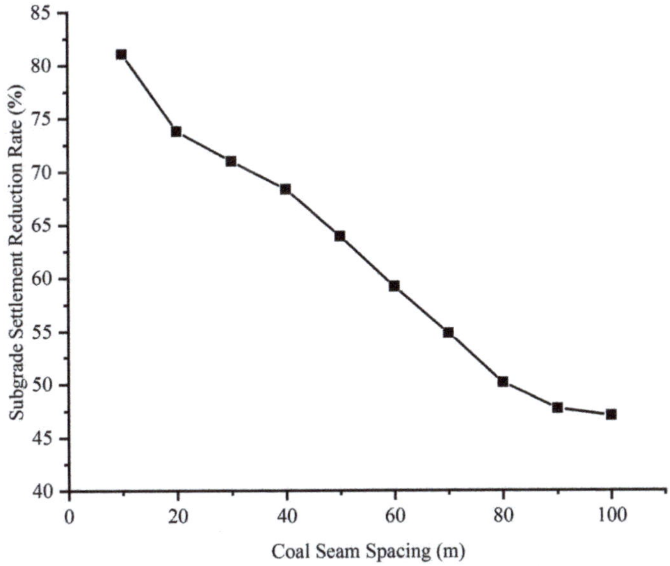

Figure 4. The subsidence reduction rate for subgrade center grouting treatment under different spacing conditions.

Figure 5 shows the relationship between the center subsidence reduction rate of the subgrade and OB/IB after grouting treatment with different coal seam spacing. From the graph, it can be concluded that when the burial depth (OB) of the upper coal seam is constant, the ratio of subgrade center subsidence reduction after grouting treatment first increases linearly ($1 \leq OB/IB \leq 2$), then increases exponentially ($2 < OB/IB$), and finally tends to be flat.

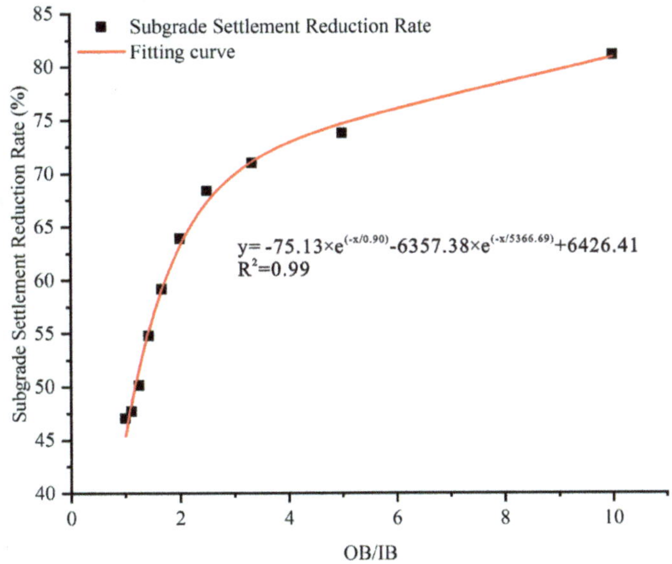

Figure 5. The subsidence reduction rate of subgrade center grouting treatment for different OB/IB conditions.

In summary, due to the large spacing between coal seams, the OB/IB is relatively small. At this time, the compaction amount of the lower coal seam caving zone is relatively large, and the upper coal seam caving zone and crack zone undergo compaction and closure under the disturbance, resulting in poor grouting treatment and a subsidence reduction effect. When the spacing between coal seams is small and OB/IB is larger, the load on the overlying rock mass is smaller, and the long-term deformation of the reinforcement after grouting is reduced, making the grouting subsidence reduction effect better.

3.2. Calculation Results of Subsidence Reduction in Goaf Areas of Multi-Coal Seam Goafs with Different Grouting Ranges

To analyze the impact of different grouting treatment ranges on the subsidence reduction effect of grouting in multi-coal seam goafs, this study considered four grouting treatment ranges: half-width range of subgrade, width range of subgrade, trapezoidal range, and full-range grouting. Figure 6 shows the residual subsidence cloud map of the goaf under different grouting treatment conditions. From the residual subsidence cloud maps of grouting, it can be concluded that under different grouting range conditions, the residual subsidence near the goaf opening and stopping line is relatively more significant than that in the mining center area. When grouting is carried out in the goaf of multi-coal seam according to the half-width of the subgrade, the width of the subgrade, and the trapezoidal range, the residual subsidence at the subgrade is controlled, and the overall residual subsidence control effect of the model after full-range grouting is the most obvious. This is because when grouting is carried out according to the width of the subgrade, half-width of the subgrade, and the trapezoidal range, the broken rock masses near the stop mining line are not treated, resulting in damage to the cantilever beam structure near the stop mining line when residual subsidence occurs, and large-scale subsidence occurs at the stop mining line. According to the full range of grouting, the damage of the cantilever beam structure near the stop mining line is controlled, forming a relatively stable structure, and the most obvious effect of reducing subsidence control is achieved [30].

Figure 7 shows the surface residual subsidence curve for different grouting treatment scopes. The results show that when using full-range grouting treatment, the residual subsidence on the surface presents a symmetrical "W"-shape on both sides. The overall trend of the fluctuations of the residual subsidence curve in the "W"-shape is relatively stable, and the best subsidence reduction effect is achieved. When grouting treatment is performed according to the entire width of the subgrade, half-width of the subgrade, and trapezoidal range, this results in damage to the cantilever beam structure near the stopping line during residual subsidence. The residual subsidence shows a tilted "W"-shape inclined toward the opening.

Figure 8 shows the amount of subsidence reduction ΔL and the ratio of R at the center of the subgrade for the range of grouting treatments. From the figure, it can be concluded that when using the full-range grouting treatment, the grouting has the best effect on reducing subsidence. At this time, the subsidence reduction amount and ratio at the center of the subgrade are 0.158 m and 81.1%, respectively. When the grouting treatment range is trapezoidal, the width of the subgrade, or half-width of the subgrade, the subsidence reduction rate at the center of the subgrade gradually decreases. Therefore, during the same period after the completion of mining, the maximum controlled residual subsidence value of the surface can be achieved when the treatment width is the full-range of the caving zone, followed by the trapezoidal range > the width of the subgrade > the half-width of the subgrade.

The inclination value and uneven subsidence control rate caused by uneven subsidence of the subgrade for different grouting treatment ranges are plotted in Figure 9. From the figure, it can be concluded that when the treatment width is selected within the range of subgrade width or half-width of the subgrade, the improvement effect on uneven subsidence of the subgrade is poor. When the treatment width is selected as trapezoidal and full-range grouting, the uneven subsidence of the subgrade is significantly improved.

After mining, the goaf is immediately treated with full-range grouting, with a control rate of over 95% for uneven subsidence of the subgrade. Trapezoidal range grouting is used with a control rate of over 56% for uneven subsidence of the subgrade.

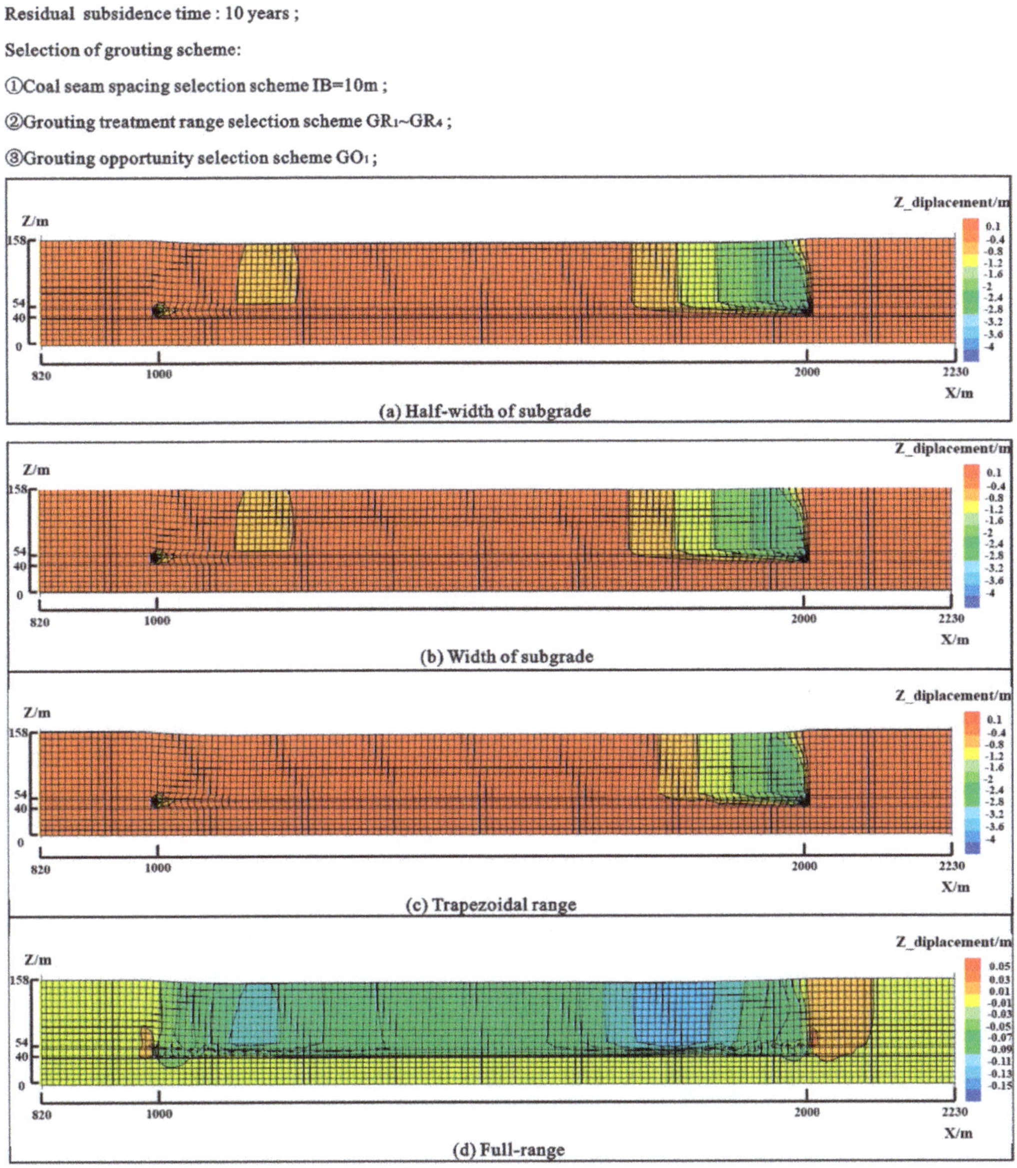

Figure 6. Residual subsidence nephogram for different grouting ranges.

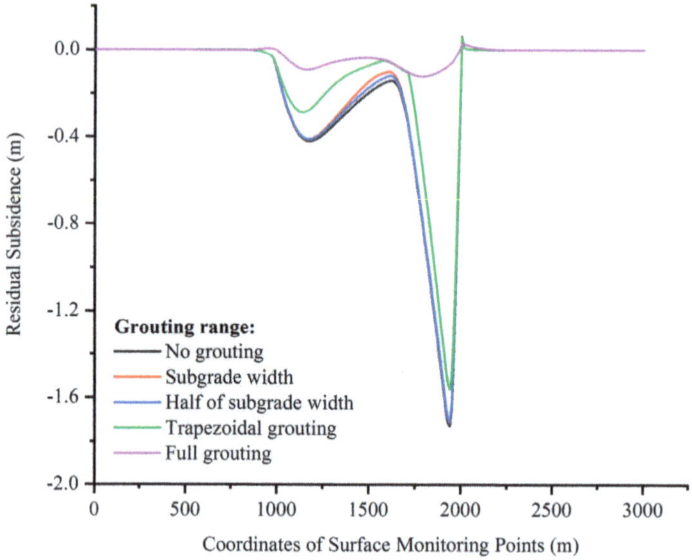

Figure 7. The surface subsidence curve for different grouting range conditions.

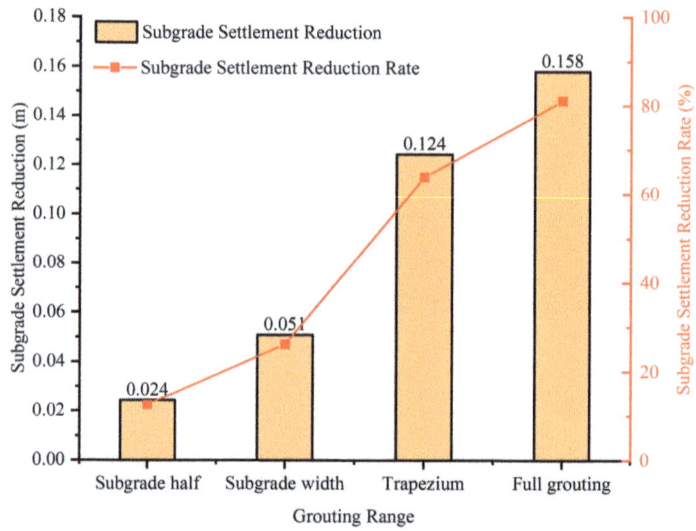

Figure 8. The amount of subsidence reduction and the subsidence reduction rate of the subgrade center for different grouting range conditions.

In summary, when the treatment width adopts the trapezoidal range and full-range grouting, the subsidence reduction at the center of the subgrade and uneven subsidence of the subgrade are significantly improved. Through simulation calculations, when the treatment width is treated according to the trapezoidal range and the control effect meets the standard requirements, the trapezoidal range can be used for treatment. The advantage is that the trapezoidal range has a smaller width than the full-range grouting treatment, and the impact on the environment after grouting treatment is relatively small, which reduces engineering costs and shortens the construction period. If the engineering and regulatory requirements are not met, then full-range grouting can be used for prevention and control, which can achieve good subsidence control effects.

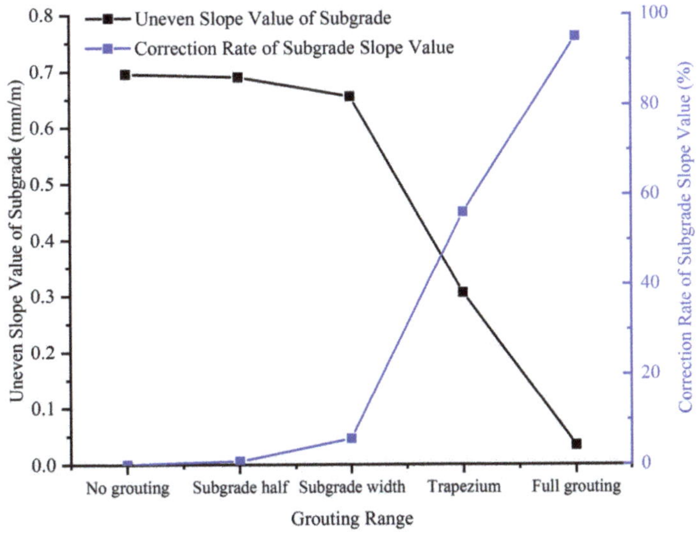

Figure 9. Subgrade incline value and incline reduction rate for different grouting range conditions.

3.3. Results of Calculations of Grouting Subsidence Reduction for Different Grouting Opportunities

This study considers conducting full-range grouting treatment within 0 to 3 years after multi-seam mining. Figure 10 shows the calculation results for the cloud maps of the residual subsidence of the goaf for different grouting opportunity conditions. The residual subsidence contour indicates grouting is carried out closer to the mining stopping on the working face, resulting in more residual subsidence in the goaf of the multi-coal seam, effectively controlling surface subsidence. When grouting is carried out after mining stops for two years, there is little difference in the subsidence contours of the goaf in the multi-coal seam. In addition, the impact of grouting opportunities on the grouting treatment effect of the goaf in the multi-coal seam is weakened, since the residual subsidence capacity of the goaf in the multi-coal seam gradually weakened after the completion of coal seam excavation. Therefore, when the residual subsidence capacity is large, grouting at a time point closer to the cessation of mining can produce better grouting and subsidence reduction effects. In summary, the later the grouting opportunity is, the greater the residual subsidence in the goaf of the multi-coal seam that has already occurred before grouting, resulting in a small amount of grouting subsidence reduction and poor grouting subsidence reduction effect.

The relationship between the subsidence reduction ratio R at the center of the subgrade and the opportunity for grouting is plotted in Figure 11. From the graph, it can be concluded that immediately after the coal seam is mined, grouting has the best effect on reducing subsidence with a subsidence reduction ratio of 67.2% at the center of the subgrade. The longer the distance from stopping mining, the subsidence reduction ratio at the center of the subgrade gradually decreases after grouting treatment, and the subsidence reduction rate curve shows an exponential relationship. When grouting is carried out after mining is stopped for 2 years with the extension of grouting opportunity, the rate of decrease in the center subsidence rate of the subgrade gradually decreases, and the curve tends to be flat.

Residual subsidence time : 10 years ;

Selection of grouting scheme :

① Coal seam spacing selection scheme IB=50m ;

② Grouting treatment range selection scheme GR₄ ;

③ Grouting opportunity selection scheme GO=0、1、2、3 year;

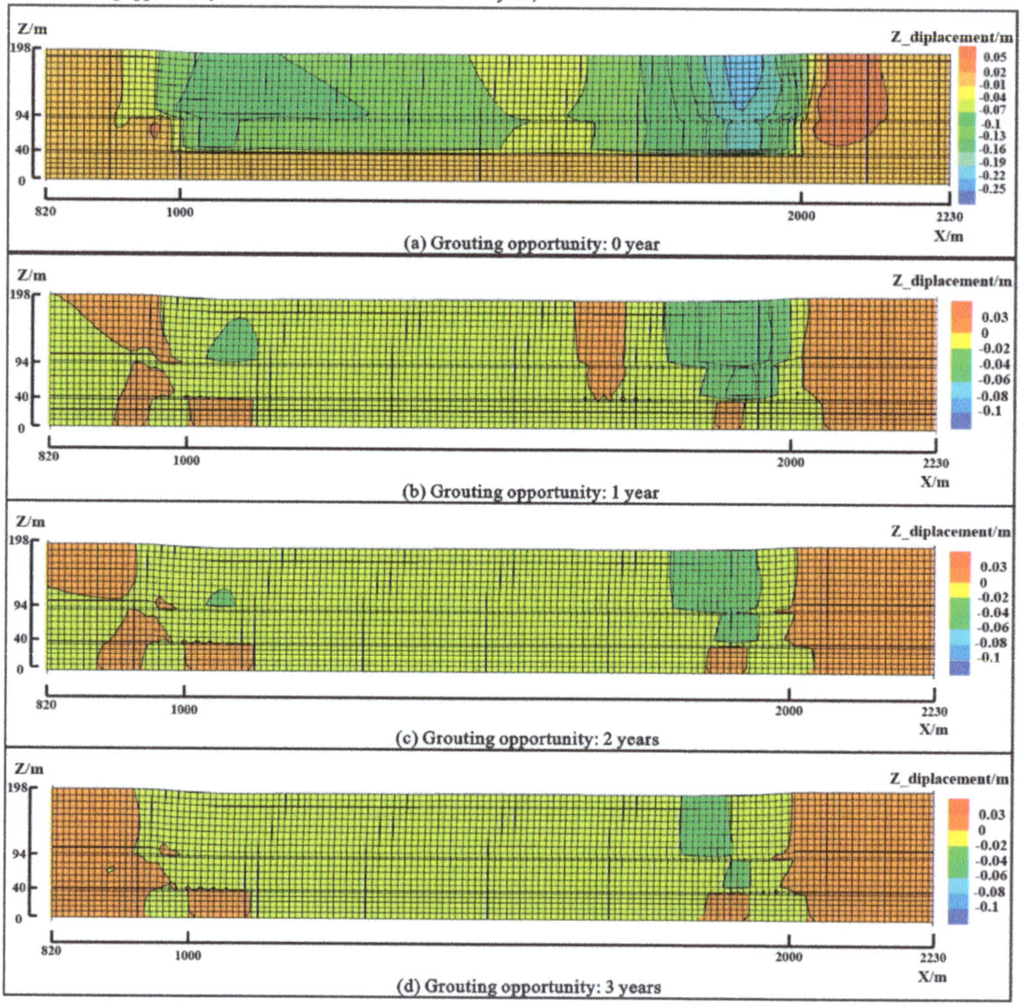

Figure 10. Residual subsidence nephogram for different grouting opportunity conditions.

In summary, due to the completion of the excavation of the working face, the surface subsidence rate gradually slows, and the residual deformation capacity gradually decreases. When the residual deformation capacity is large then grouting is performed closer to the completion of the excavation of the working face, which can produce better grouting and subsidence reduction effects.

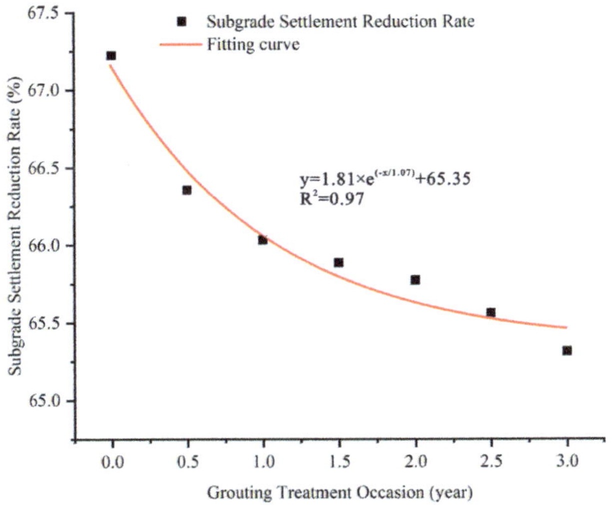

Figure 11. Subgrade center subsidence reduction rate for different grouting opportunity conditions.

Figure 12 shows the variation of the center subsidence reduction ratio of the subgrade for different grouting opportunities and coal seam spacing conditions. From the figure, it can be seen that with the extension of grouting opportunity and the decrease in OB/IB, the subsidence reduction ratio at the center of the subgrade decreases exponentially. When $1 \leq OB/IB \leq 2$ and grouting is carried out 2 years after excavation, the subsidence reduction ratio of the subgrade center decreases rapidly, and the grouting treatment effect is poor. This is because increasing the spacing between coal seams can lead to increased deformation of the surface and goaf, and even with grouting treatment, it is difficult to completely prevent the problem of surface subsidence. With the extension of grouting opportunity, there is a significant amount of residual deformation in the goaf, resulting in a decrease in the contact area between grouting materials and fractured rock masses and cracks in the goaf, which influences the grouting effect.

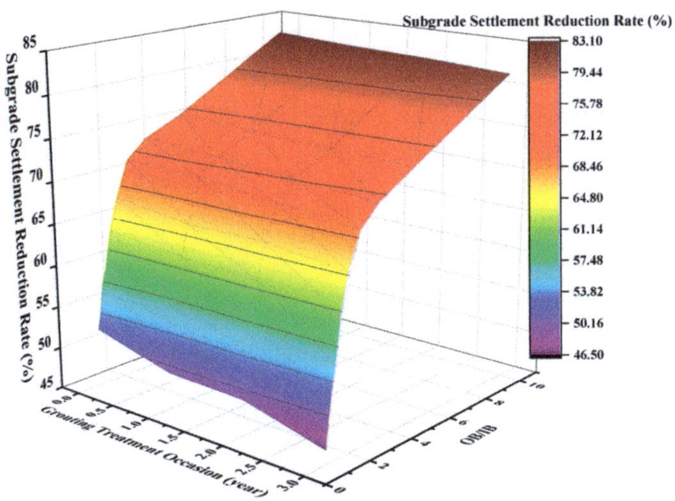

Figure 12. Diagram of the relationship between grouting opportunity, OB/IB, and subgrade center subsidence reduction rate.

4. Conclusions

Currently, there is no dependable foundation for constructing grouting treatment plans for multi-seam goafs. This study adopted a numerical simulation method to design models of multi-coal seam goafs with different spacing conditions. By changing the grouting range and opportunity, the influence of coal seam spacing, grouting range, and grouting opportunity on the subsidence reduction effect of multi-coal seam goafs was studied. The main conclusions are as follows:

(1) With constant overburden (OB), the subsidence reduction ratio of the subgrade center increases linearly as the interburden (IB) decreases ($1 \leq OB/IB \leq 2$), then increases exponentially ($2 < OB/IB$), and eventually becomes stable.

(2) When the treatment is performed according to the half-width of the subgrade, the width of the subgrade and the range of a trapezoid, the surface residual subsidence tends to the inclined 'W'-shape of the open cutting. The surface residual subsidence has a symmetrical 'W'-shape when full-range grouting is adopted. When the treatment width is selected as the width of the subgrade or half-width of the subgrade, the improvement effect on uneven subsidence of the subgrade is poor. When the treatment width is selected as the trapezoidal range and full-range grouting, the uneven subsidence of the subgrade is significantly improved.

(3) For the grouting treatment of goaf areas with multi-coal seam goafs where mining has been stopped for a longer time, the central subsidence reduction rate of the subgrade is smaller, and there is an exponential relationship between the central subsidence reduction rate of the subgrade and the grouting opportunity. With the extension of grouting opportunity and the decrease in OB/IB, the subsidence reduction ratio at the center of the subgrade decreases in an exponential curve.

In the simulation process of this article, the results of large-scale mining cases under real overburden were not verified, which requires further improvement based on the simulation in this article.

Author Contributions: Conceptualization, H.W. and Y.Q.; methodology, Y.Q. and Y.G.; software, H.W. and Y.G.; validation, Y.Q. and N.X.; formal analysis, Y.G. and Y.Q.; investigation, H.W. and Y.Q.; resources, Y.Q. and N.X.; data curation, Y.G. and Y.Q.; writing—original draft preparation, Y.G. and Y.Q.; writing—review and editing, Y.Q. and N.X.; visualization, H.W. and Y.G.; supervision, Y.Q. and N.X.; project administration, Y.Q.; funding acquisition, N.X. All authors have read and agreed to the published version of the manuscript.

Funding: This research was funded by National Natural Science Foundation of China (NSFC) under grant no. 42230709 with the title "Mechanism and prevention methods of tunnel disaster induced by long-term deformation of rock strata in mining-induced subsidence zone".

Data Availability Statement: Not applicable.

Acknowledgments: The authors thank the National Natural Science Foundation of China (NSFC) under Grant No. 42230709.

Conflicts of Interest: The authors declare no conflict of interest.

References

1. Wang, H.; Li, Y.; Dun, Z.; Cheng, J.; Dun, Z.; Wu, C. Seepage Grouting Mechanism for Foundations in Goaf Sites considering Diffusion Paths. *Geofluids* **2022**, *2022*, 8394811. [CrossRef]
2. Wei, X.; Bai, H.; Rong, H.; Jiao, Y.; Zhang, B. Research on Mining Fracture of Overburden in Close Distance Multi-seam. *Procedia Earth Planet. Sci.* **2011**, *2*, 20–27. [CrossRef]
3. Zhang, M.; Shimada, H.; Sasaoka, T.; Matsui, K.; Dou, L. Evolution and effect of the stress concentration and rock failure in the deep multi-seam coal mining. *Environ. Earth Sci.* **2013**, *72*, 629–643. [CrossRef]
4. Adhikary, D.; Khanal, M.; Jayasundara, C.; Balusu, R. Deficiencies in 2D Simulation: A Comparative Study of 2D Versus 3D Simulation of Multi-seam Longwall Mining. *Rock Mech. Rock Eng.* **2016**, *49*, 2181–2185. [CrossRef]

5. Ma, W.-M.; Zhu, W.-Y. Effect of multi-seam mining on subsidence. *Int. J. Min. Eng.* **1984**, *2*, 171–173. [CrossRef]
6. Ren, G.; Li, G.; Kulessa, M. Application of a Generalised Influence Function Method for Subsidence Prediction in Multi-seam Longwall Extraction. *Geotech. Geol. Eng.* **2014**, *32*, 1123–1131. [CrossRef]
7. Ma, W.J.; Hu, H.F. The Strata Movement Regularity and Parameter Simulation Analysis in Multiple Seams Repeated Mining. *Appl. Mech. Mater.* **2013**, *295–298*, 2935–2939. [CrossRef]
8. Ying-Ke, L.; Fu-Bao, Z.; Lang, L.; Chun, L.; Shen-Yong, H. An experimental and numerical investigation on the deformation of overlying coal seams above double-seam extraction for controlling coal mine methane emissions. *Int. J. Coal Geol.* **2011**, *87*, 139–149. [CrossRef]
9. Ghabraie, B.; Ren, G.; Smith, J.V. Characterising the multi-seam subsidence due to varying mining configuration, insights from physical modelling. *Int. J. Rock Mech. Min. Sci.* **2017**, *93*, 269–279. [CrossRef]
10. Ghabraie, B.; Ren, G.; Barbato, J.; Smith, J.V. A predictive methodology for multi-seam mining induced subsidence. *Int. J. Rock Mech. Min. Sci.* **2017**, *93*, 280–294. [CrossRef]
11. Salmi, E.F.; Nazem, M.; Deng, K.Z.; Karakus, M. Numerical analysis of a large landslide induced by coal mining subsidence. *Eng. Geol.* **2017**, *217*, 141–152. [CrossRef]
12. Bin Zhang, X. The Probe Technology and Counter Measures of Goaf of Colliery Area in Tongluoshan Tunnel. *Appl. Mech. Mater.* **2012**, *204–208*, 1419–1422. [CrossRef]
13. Zha, W.H. Reinforced Techniques Analysis of Goaf-Side Entry at Isolated Island Coal Face Based on FLAC Numerical Simulation. *Adv. Mater. Res.* **2012**, *605–607*, 210–214. [CrossRef]
14. Cai, W.; Chang, Z.; Zhang, D.; Wang, X.; Cao, W.; Zhou, Y. Roof filling control technology and application to mine roadway damage in small pit goaf. *Int. J. Min. Sci. Technol.* **2018**, *29*, 477–482. [CrossRef]
15. Xie, S.; Pan, H.; Zeng, J.; Wang, E.; Chen, D.; Zhang, T.; Peng, X.; Yang, J.; Chen, F.; Qiao, S. A case study on control technology of surrounding rock of a large section chamber under a 1200-m deep goaf in Xingdong coal mine, China. *Eng. Fail. Anal.* **2019**, *104*, 112–125. [CrossRef]
16. Bai, H.R.; Li, J.J. Evaluation of Grouting Effect Detection in Goaf. *Appl. Mech. Mater.* **2013**, *438–439*, 1080–1083. [CrossRef]
17. Ao, X.; Wang, X.; Zhu, X.; Zhou, Z.; Zhang, X. Grouting Simulation and Stability Analysis of Coal Mine Goaf Considering Hydromechanical Coupling. *J. Comput. Civ. Eng.* **2017**, *31*, 04016069. [CrossRef]
18. Wu, X.; Wang, S.; Tian, C.; Ji, C.; Wang, J. Failure Mechanism and Stability Control of Surrounding Rock of Docking Roadway under Multiple Dynamic Pressures in Extrathick Coal Seam. *Geofluids* **2020**, *2020*, 8871925. [CrossRef]
19. Li, X.; Guo, X.; Sun, G. Grouting Reinforcement Mechanism and Multimodel Simulation Analysis of Longwall Goaf. *Geofluids* **2021**, *2021*, 13. [CrossRef]
20. Hu, Z.; Ma, B.; Chen, X.; Chen, L. Study on Sensitivity Parameters Analysis of Grouting Reinforcement Underpassing Existing Subway Tunnel by Numerical Modeling. *Adv. Civ. Eng.* **2021**, *2021*, 13. [CrossRef]
21. Xuan, D.; Li, J.; Zheng, K.; Xu, J. Experimental Study of Slurry Flow in Mining-Induced Fractures during Longwall Overburden Grout Injection. *Geofluids* **2020**, *2020*, 8877616. [CrossRef]
22. Yu, W.; Li, K.; Liu, Z.; An, B.; Wang, P.; Wu, H. Mechanical characteristics and deformation control of surrounding rock in weakly cemented siltstone. *Environ. Earth Sci.* **2021**, *80*, 337. [CrossRef]
23. Liu, B.; Sang, H.; Wang, Z.; Kang, Y. Experimental Study on the Mechanical Properties of Rock Fracture after Grouting Reinforcement. *Energies* **2020**, *13*, 4814. [CrossRef]
24. Xie, S.; Wu, X.; Chen, D.; Sun, Y.; Wang, E.; Wu, X.; Duan, X.; Jiang, Z.; Qi, P.; Shi, S. Failure Mechanism and Control Technology of Thick and Soft Coal Fully Mechanized Caving Roadway under Double Gobs in Close Coal Seams. *Shock. Vib.* **2020**, *2020*, 8846014. [CrossRef]
25. Han, L.J.; He, Y.N. Experimental Study on Mechanical Characteristics of Cracked Rock Mass Reinforced by Bolting and Grouting. *J. China Univ. Min. Technol.* **2005**, *34*, 418–422.
26. Modoni, G.; Bzówka, J. Analysis of Foundations Reinforced with Jet Grouting. *J. Geotech. Geoenvironment. Eng.* **2012**, *138*, 1442–1454. [CrossRef]
27. Zong, Y.; Han, L.; Qu, T.; Yang, S. Mechanical properties and failure characteristics of fractured sandstone with grouting and anchorage. *Int. J. Min. Sci. Technol.* **2014**, *24*, 165–170. [CrossRef]
28. Wang, C.; Xiong, Z.; Wang, C.; Wang, Y.; Zhang, Y. Study on Rib Sloughage Prevention Based on Geological Structure Exploration and Deep Borehole Grouting in Front Abutment Zones. *Geofluids* **2020**, *2020*, 7961032. [CrossRef]
29. Deng, K.Z.; Tan, Z.X.; Zhang, H.Z. Design Method of Strip Grouting for Old Longwall Mining Goaf. *J. China Coal Soc.* **2008**, *161*, 153–156.
30. Chen, S.J.; Zhang, L.B.; Jiang, N.; Yin, D.; Gao, Z.; Guo, W. A case of large buildings construction above oldmine goaf in Shandong Province. *J. China Coal. Soc.* **2022**, *47*, 1017–1030.
31. He, F.; Lv, K.; Li, X.; Qin, B.; Li, L. Failure Mechanism and Control of Lower Retracement Channel in Close-Distance Double-Thick Coal Seams. *Shock. Vib.* **2021**, *2021*, 6651099. [CrossRef]

32. Shanxi Traff Layout Reconnaissance Design Inst Co. JTG/T D31-03. *Guidelines for Design and Construction of Highway Engineering in the Mined-Out Area*; People's Communications Publishing House: Beijing, China, 2011.
33. Qin, Y. *Restudy on Complex Mechanical Behavior of Rock Mass in Multi-Seam Mining*; China University of Geosciences: Beijing, China, 2021.
34. Yue, Z. *Simulation Study on Long-Term Deformation Law of Coal Mine Goaf Based on FLAC3D*; China University of Geosciences: Beijing, China, 2023.

applied sciences

MDPI

Article

Effects of the Excavation of a Hydraulic Tunnel on Groundwater at the Wuyue Pumped Storage Power Station

Tong Jiang [1,*], Xun Pei [2], Wenxue Wang [3], Longfei Li [2] and Shihao Guo [2]

1 Henan Joint International Research Laboratory of Collapse-Landslide-Debris Flow Monitoring and Early Warning, North China University of Water Resources and Electric Power, Zhengzhou 450046, China
2 College of Geosciences and Engineering, North China University of Water Resources and Electric Power, Zhengzhou 450046, China; b2020081811@stu.ncwu.edu.cn (X.P.)
3 Henan Key Laboratory Rock & Soil Mech & Struct Engineering, North China University of Water Resources and Electric Power, Zhengzhou 450046, China
* Correspondence: jiangtong@ncwu.edu.cn

Abstract: The tailwater tunnel of the Wuyue pumped storage power station is located in bedrock and extends to depths between tens and hundreds of meters. It is impossible to analyze and evaluate the whole engineering area from geological exploration data, and the hydrogeological conditions are complicated. In the early stages of the tailwater tunnel's construction, the drinking water wells in four villages dried up. This paper reports the results from a field investigation, in situ tests, laboratory tests, and numerical simulation carried out to determine how the groundwater was affected when the tunnel was excavated. A hydrogeological model of the region was established from the inverted regional natural flow field parameters. The model was validated, and an analysis of the errors showed that there was an average error of 1.98% between the natural flow field and the hydrogeological survey flow field. The model was then used to simulate the three-dimensional transient seepage fields under normal seepage conditions and limited seepage conditions, as far as was practical. The results showed that, as the excavation of the tailwater tunnel advanced, the water inflow to the tunnel also increased. When the water inflow increased from 1000 to 5000 m^3/d, the water level at a distance of 100 m from the axis of the tunnel dropped from −0.956 to −1.604 m. We then analyzed how the water level changed as the water inflow varied and proposed a formula for calculating the extent of the influence on the groundwater. We studied how the water level changed at different well points to ascertain how a groundwater well became depleted and determined the factors that influenced seepage in the regional flow field. The water level in different areas of the project area was simulated and analyzed, and the extent of the groundwater area affected by the tunnel construction was clarified. We then studied how the groundwater in different areas of, and distances from, the project area was influenced by normal seepage conditions and limited seepage conditions and proposed a formula for calculating the extent of the influence on groundwater for different water inflows. We constructed a 'smart site' for visualizing data, sharing information, and managing the project. Time–frequency domain analysis was applied to explore the extent of the impacts and range of the vibration effects on residential housing at different distances from the project area caused by the different methods for excavating the tailwater tunnel. The results from this analysis will provide useful insights into how the excavation of this tailwater tunnel will impact the local residents and living areas.

Keywords: pumped storage; Wuyue Reservoir; transient numerical simulation; parameter inversion; seepage field

check for updates

Citation: Jiang, T.; Pei, X.; Wang, W.; Li, L.; Guo, S. Effects of the Excavation of a Hydraulic Tunnel on Groundwater at the Wuyue Pumped Storage Power Station. *Appl. Sci.* **2023**, *13*, 5196. https://doi.org/10.3390/app13085196

Academic Editor: Joao Carlos Andrade dos Santos

Received: 22 March 2023
Revised: 18 April 2023
Accepted: 19 April 2023
Published: 21 April 2023

1. Introduction

In recent years, a large number of hydraulic tunnels have been built in China. While these hydraulic tunnels help to promote social progress, they also destroy the groundwater environment. When hydraulic tunnels are being excavated, a large volume of groundwater is lost, and the groundwater levels continuously decline, causing problems such as exhaustion of surface wells and springs, groundwater leakages, and ground collapse. Groundwater leakage reduces the soil water content, with serious consequences for the ecological environment, such as decreases in, or even the disappearance of, regional vegetation [1]. It is, therefore, very important to study how seepage fields of hydraulic tunnels evolve during construction and how groundwater levels change because of seepage around the tunnel.

Numerous researchers have studied how tunnel construction affects the groundwater environment [2–10]. Underground engineering and construction disturb the groundwater environment and lead to decreases in the water resources across wide geographical areas and over considerable time periods; in some cases, the effects are irreversible and the groundwater does not recover [11–15]. When precipitation events occur during the construction of a tunnel, large amounts of groundwater are discharged from natural aquifers within a certain spatial range leading to reductions in the underground water levels and local changes in the groundwater storage characteristics, such as the evolution of landing funnels and changes in groundwater seepage fields. Given that these changes might compromise groundwater supplies, algorithms of underground water inrush into such tunnels have therefore been proposed to predict how the groundwater environment would be affected by the construction of a tunnel [16,17]. Researchers have studied how the groundwater flow changes when rock tunnels are excavated and have found that groundwater circulation systems may be damaged by tunnel water gushing during excavation [18–22]. The surface water source and some spring points may dry up during water flushing accidents in the construction phase. Studies to date, therefore, have shown that underground engineering construction can have serious effects on the groundwater environment and the supply of groundwater resources [10,23,24].

During the construction of an ultra-long subway tunnel in Seoul, South Korea, in 2008, the groundwater level dropped dramatically in some parts of the project area, and a large number of underground wells dried up and were abandoned. Calculations showed that the average annual water yield reached 6.3×10^7 m^3, and the local groundwater environment was severely damaged [25]. During the construction of a 7300 m tunnel on the high-speed railway line between Malaga and Cordoba in a karst area of southern Spain, there was a large water inrush with an instantaneous flow of up to 800 L/s, which resulted in serious shortages of groundwater supplies for drinking and irrigation in that area [26]. In 2009, because of a drainage problem in the Frizenzuola Tunnel, Italy, five streams in the tunnel site were cut off and 12 springs dried up [27]. Japanese scholars studied the groundwater levels during the construction of a tunnel in Matsumoto City, Japan. They found that a large amount of groundwater gushed into the tunnel, leading to groundwater leakages, exhaustion of springs within the scope of its influence, and decreases in the river flow in the discharge area [28]. In 2010, the Yujiawan Reservoir dried up during the construction of the Xiyong Tunnel in Chongqing, thereby damaging the function and production ability of approximately 50 mu of farmland in Jingang Village. In March 2012, drinking water sources were depleted in Shiyuan Village, Zhongliang Town, during the construction of the Zhongliangshan Tunnel on the Lanzhou-Chongqing Railway, and more than 1300 people and 10 businesses experienced severe water shortages.

If relevant measures are not taken to control the discharge of groundwater during the construction of a tunnel, groundwater levels may decline dramatically, causing drainage of surface water within a certain range and groundwater to a certain depth. The water supplies, such as springs and municipal wells, within the range of the construction will decrease to varying degrees, with serious impacts on the lives and activities of local residents.

The Wuyue pumped storage power station in Henan Province, China, is located in the transition zone between a plain area and a mountainous area. The rivers in the study area belong to the Huaihe River system. The overall aim of this study was to reduce the impact of a tunnel construction project on the water resources and lives of residents around the construction area by gaining an accurate understanding of how groundwater resources would change during the construction period. Based on the systematic analysis of the hydrogeological characteristics and key geological structures of the area around the Wuyue pumped storage power station, a conceptual model of the hydrogeology in the area is developed. A three-dimensional unsteady seepage numerical model of groundwater that reflected the three-dimensional hydrogeological structure and unit partition of the area was established. The rate of the model convergence was studied, and then the three-dimensional unsteady seepage in the area around the Wuyue pumped storage power station was numerically simulated based on the finite element method. The seepage field of the project area under natural conditions and five possible seepage control design schemes for the tunnel construction were simulated and analyzed. From this analysis, a seepage control design scheme was identified that was approximately consistent with the anticipated water inflows during the project. This control scheme can be used as the basis for calculating and analyzing changes in the seepage and groundwater in the tailwater tunnel in a later phase of the construction.

2. Engineering Materials and Methods

2.1. Overview of the Study Area

The Wuyue pumped storage power station is located in Yinpeng Township, Guangshan County, Henan Province, China. The straight-line distances between the Wuyue pumped storage power station and Zhengzhou and Xinyang are 340 and 70 km, respectively, and the Wuyue pumped storage power station is 42 km from Guangshan County by highway (Figure 1), meaning that it is relatively easy to reach by normal transport. The project area comprises eroded low hills and gullies that are characteristic of southern Henan. The river valleys are well-developed, and the area comprises low mountains, hills, loess ridges, and river plains from south to north. The upper reservoir of the power station is in the Laoshanzhai Mountain Forest area, close to the top of the north slope of Laoshanzhai. The reservoir has three large main branches, and the power station is located at the tail of the western branch (Figure 2). A three-dimensional digital elevation model of the engineering area is shown in Figure 2.

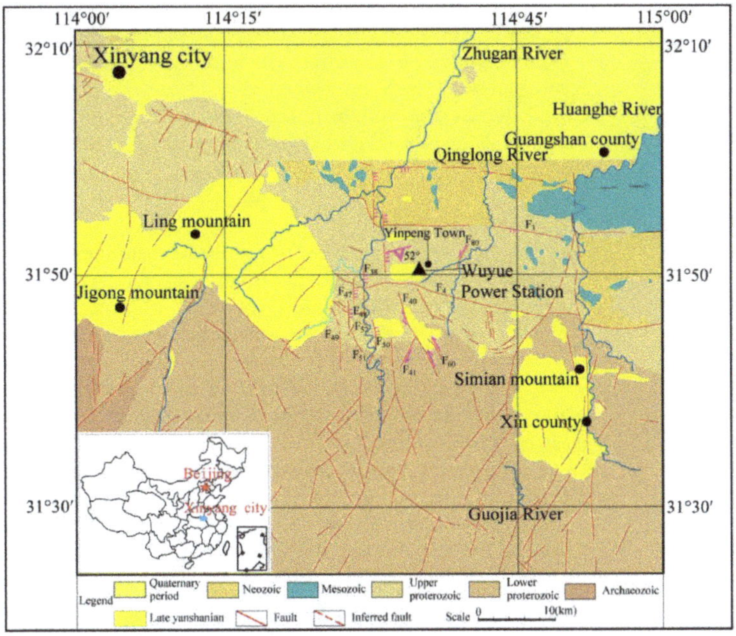

Figure 1. Regional geological structure of the area close to the Wuyue power station.

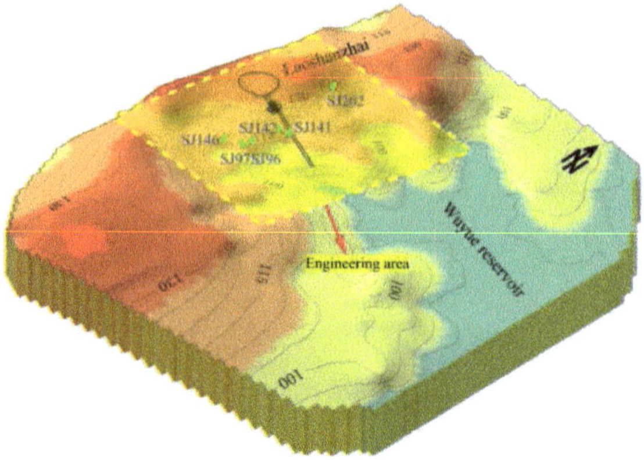

Figure 2. A three-dimensional digital elevation model of the engineering area.

The tailwater tunnel is arranged with two holes and four machines. The two main tailwater holes (diameter 8.0 m) are circular and are arranged parallel to the pressure holes at an axis orientation of N95.5° E, and the distance between the axes is 46.0 m. The design outlet floor is at an elevation of 60.0 m, the tailwater tunnel is 1456.5 m long, and the buried depth varies from 20 to 285 m. The rock surrounding the shallow tunnel is weak and severely weathered, and the rock structure is broken. The construction of a large-diameter underground cavern has resulted in considerable disturbance and destruction to a large area of rock surrounding the cavern. Therefore, the type, structure, and degree of weathering of the surrounding rock have seriously restricted the engineering design and construction.

A hydrogeological survey was carried out in the early stage of the tailwater tunnel construction to measure the water levels of the wells and ponds in the region. The results from this survey showed that there was a positive linear relationship between the groundwater level and the surface elevation, that the water level of the shallow wells was controlled by the surface elevation, and that the water was mainly from atmospheric precipitation. The hydrogeological survey results also showed that various drinking water wells were exhausted, including SJ146 in Tulou, SJ141 and SJ142 in Qiandawan, SJ96 and SJ97 in Yangshuwan, and SJ202 in Yangmingao, thereby affecting the lives of the villagers (Figure 2).

2.2. Regional Geological Setting

The project area is near the granite intrusion of the late Yanshanian and comprises the middle Proterozoic Guishan Formation schist, the Devonian Nanwan Formation granites and schist, the late Yanshanian intrusive mid-grained granites and dikes, and unevenly distributed assorted loose deposits from the Quaternary. The geological structure is mainly characterized by small-scale faults, joint fissures, and schistosity developed in the lower reservoir area. The developed fault structure has three main directions, namely NE, NNE, and NNW. The degree of development and direction of the fault structure vary throughout the engineered area.

The results from a geological survey showed that the rock surrounding the tailwater tunnel was mainly granite and metamorphic schist, that the contact zone of the two lithologies had a horizontal width of 95 m, and that the altered rock in the contact zone was poorly structured. The underlying bedrock is metamorphic schist of the Nanwan Formation with a schistosity of 273°–287°/SW < 65°–75°. There are 20 faults of different sizes in the engineering area, all of which have steep dip angles. Of these, three faults (F4, F5, and F6) in the tailwater tunnel strike in a NNE direction, with an occurrence of 6°–10°/NW and SE < 70°–77°, and faults F4 and F5 are larger than F6 (Figure 3).

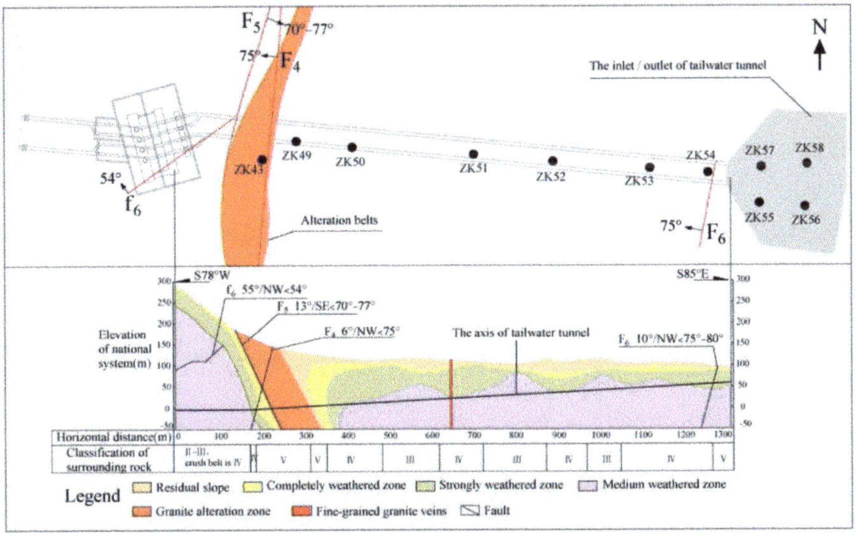

Figure 3. Geological structure and drilling layout of engineering area.

In pile numbers W①0 + 164–W①0 + 256 and W②0 + 188–W②0 + 282 of the tailwater tunnel, there is a developed granite alteration zone with a horizontal width of approximately 95 m and a metamorphic rock area with well-developed schistotectonics. The weathering characteristics of the rock mass in the engineering area, defined by indexes such as the fracture spacing, rock quality index, p-wave velocity, and RQD, were considered when choosing the arrangement of the drilling holes (Figure 3). The lower limit of the rock weathering exposed by the drilling of the tailwater tunnel is shown in Table 1.

Table 1. Summary statistics of the weathered depths of the drilled holes in the tailwater tunnel.

Engineering Position	Boring Number	Orifice Height (m)	Deep Hole (m)	Coating Thickness (m)	The Burial Depth of Lower Limit of Each Weathering Zone (m)			Note
					Complete Weathering	Strong Weathering	Medium Weathering	
Tailwater tunnel	ZK43	147.75	140.1	0	122.3	140.1	—	Alteration zone
	ZK49	131.6	90.15	9.6	90.15	—	—	Alteration zone
	ZK50	114.31	100.3	7.2	11.6	24.2	—	Complete weathering and weak weathering alternate
	ZK51	104.82	80.15	0.8	5.6	7.65	66.1	The hole depth of 48.6–59.9 m shows strong weathering
Tailwater tunnel inlet/outlet	ZK52	114.63	70.5	7	30.3	43.6	70.5	—
	ZK53	104.84	55.6	3.5	9.2	24.4	33.4	—
	ZK54	96.48	45.4	6.3	21.4	28.1	45.4	—
	ZK55	91.9	45.4	3.3	13.5	14.8	40.2	—
	ZK56	102.22	35.5	7.5	16.8	21.9	32.5	—
	ZK57	93.93	45.7	3.8	8.1	12.6	35.5	The hole depth of 35.5–45.2m shows strong weathering
	ZK58	89.78	35.4	4.7	10.3	17.4	35.4	—

The exposed rock near the granite alteration zone along the tailwater tunnel has a relatively large burial depth in the completely weathered and strongly weathered zones because of the influence of the granite intrusion, and the degree of weathering increases as the distance to the alteration zone decreases (Figure 3). The lower limits of the completely weathered zone and strongly weathered zone are at 8–25 and 12–33 m, respectively. The lower limit of the medium-weathered zone is 33–50 m deep, and the maximum depth is 70.5 m (ZK52). The completely weathered depths of ZK43 and ZK49 near the alteration zone are at 122.33 and 190.15 m, respectively.

2.3. Regional Hydrology

The project area is on the southern bank of the Huaihe River, and the rivers in the area are all part of the Huaihe River system. The main channel of the Huaihe River snakes through the northern part of the project area from west to east. The main tributaries in the area close to the southern bank include the Zhugan River, Qinglong River (the main channel of the upper reaches of the Zhai River), and the Huanghe River (Figure 1); of these, the Zhugan River is the largest tributary in the area. The Zhaihe River, a first-order tributary of the Huaihe River, originates at Huangmaolao in Qianjin Township, Xinxian County, and flows from south to north through Guangshan County and Xixian County and into the Huaihe River at Baozikou, Huangchuan County. The river is 110 km long and has a basin area of 710 km^2 and a dense network of tributaries. The Wuyue pumped storage power station is on the Qinglong River, at an elevation of 565.60 m.

2.3.1. Groundwater Flow Field

During the excavation of the tunnel, groundwater flows into the tunnel, and the water level falls, causing a decrease in the surface water level and depletion of springs and wells. This problem can be solved by determining the boundary position of the seepage field, especially the boundary of the seepage-free surface. The tailwater tunnel of the Wuyue pumped storage power station is at the tail of the Wuyue Reservoir, and the hydrogeological conditions are complex.

A hydrogeological investigation was carried out around the axis of the tailwater tunnel using a radiation method to determine the boundary position of the groundwater flow field and the boundary of the seepage-free surface. The northern and southern sides of the survey area were bounded by ridges, the western side was bounded by Laoshanzhai, and the eastern side was bounded by the Wuyue Reservoir, and the survey area covered approximately 70 km^2. The survey included 349 wells, 225 ponds, and 6 reservoirs. After the survey data were summarized, collated, and analyzed, the regional water level contour map was drawn with the Surfer software package (Figure 4). The results from the hydrogeological investigation and borehole exposure showed that the groundwater types in the engineering area were controlled by the characteristics of the stratigraphic lithology and geological structure. There were two main types of groundwater in the engineering area, namely bedrock fissure water and phreatic pore water in the upper soil. There was also a small amount of fissure-confined water in the deeply weathered rock mass or in thicker parts of the Quaternary system.

Figure 5 shows the groundwater level along the A–A′ transect (Figure 4), the elevation information of the area, and the location and elevation of the tailwater tunnel section.

2.3.2. Seepage Field in the Engineering Area

The results of the regional hydrogeological survey show that the water level in the survey area decreased gradually from west to east, with maximum and minimum water levels of 131.7 and 88 m, respectively (Wuyue Reservoir water level). The water level was at 97 m at the inlet/outlet of the tailwater tunnel, and it increased to 122 m going westward along the axis of the tailwater tunnel (Figure 5). The water level was higher on the western side of the tailwater tunnel than on the eastern side, and the water level on the northern side of the axis of the tunnel was higher than on the southern side (Figure 4). In Figure 5, curve BB′ refers to the topographic elevation change along the tailwater tunnel, and the geological structure and fault distribution in the curve BB′ section are shown in Figure 3.

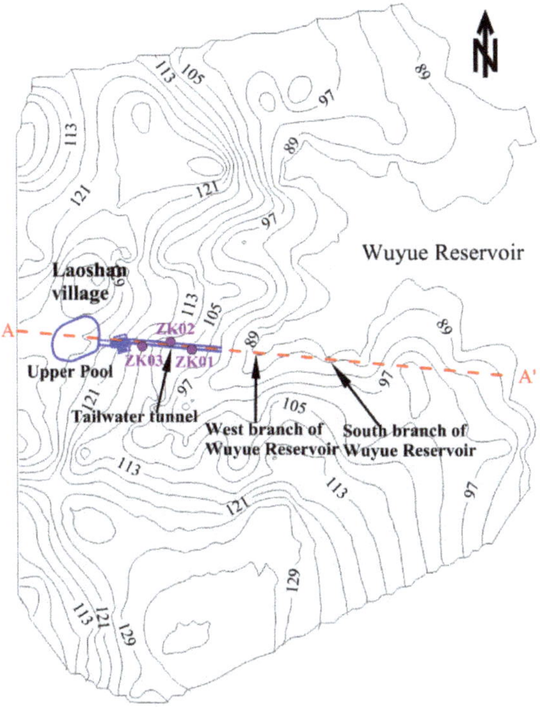

Figure 4. The scope of the hydrogeological survey and the water level isoline.

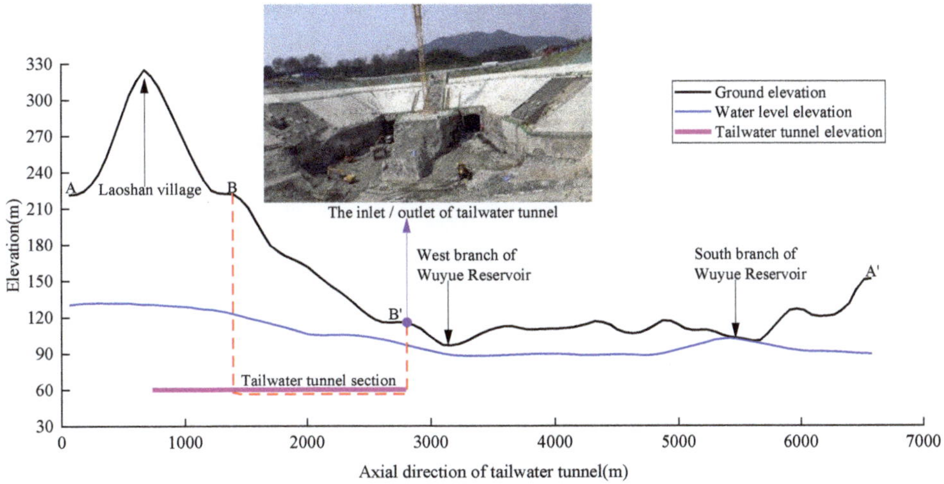

Figure 5. Groundwater occurrence along the A–A′ transect.

3. Numerical Calculation

3.1. Selection of Calculation Methods

The groundwater type is controlled by the properties of the stratum lithology and the geological structure. There are two main types of groundwater in the engineering area, namely bedrock fissure water and pore water in the overlying soil.

The pore water that exists in the slope deposits and residues is mainly distributed in the surface residual slope deposits and alluvial flood deposits with large gullies, and it is relatively uniformly distributed and well-connected hydraulically. The groundwater in the whole project area is dominated by the bedrock fissure type. Because of the different causes and distribution of cracks in the rock mass, there are non-uniform fissures throughout the bedrock aquifer, and bedrock fissure water develops. Under the combined effects of seepage and stress, the hydrogeological conditions of the aquifer around the tunnel will change because of the engineering work, and the seepage channels of the fractured rock mass will be dredged, meaning the aquifer permeability will increase.

In the three-dimensional seepage analysis, the rock mass is regarded as an equivalent and continuous anisotropic medium, and seepage is considered to occur mainly in the rock mass. Fracture medium seepage is commonly studied by the equivalent continuous medium seepage model and double permeability medium model. Because there were no data about the distribution of cracks in the engineering area, the double permeability medium model could not be used, so the equivalent continuous medium seepage model can be used. Given the engineering geological and hydrogeological conditions of the tailwater tunnel, the aim of the modeling was to determine the hydrogeological unit partition, seepage characteristics of each stratum, and the tunnel boundary conditions. The three-dimensional transient simulation of groundwater seepage in the tailwater tunnel was determined using the finite element method of the equivalent continuum seepage model.

3.2. Differential Equation of Seepage Analysis

Many researchers [29–31] have studied groundwater seepage in tunnel engineering. The differential equation that governs three-dimensional seepage in a rock mass is expressed as:

$$\text{Basic equation of seepage}: \frac{\partial}{\partial x}\left(k_x\frac{\partial H}{\partial x}\right)+\frac{\partial}{\partial y}\left(k_y\frac{\partial H}{\partial y}\right)+\frac{\partial}{\partial z}\left(k_z\frac{\partial H}{\partial z}\right)+Q=\frac{\partial \theta}{\partial t}, \quad (1)$$

$$\text{Steady flow equation}: \frac{\partial}{\partial x}\left(k_x\frac{\partial H}{\partial x}\right)+\frac{\partial}{\partial y}\left(k_y\frac{\partial H}{\partial y}\right)+\frac{\partial}{\partial z}\left(k_z\frac{\partial H}{\partial z}\right)+Q=0, \quad (2)$$

and

$$\text{Unsteady flow equation}: \frac{\partial}{\partial x}\left(k_x\frac{\partial H}{\partial x}\right)+\frac{\partial}{\partial y}\left(k_y\frac{\partial H}{\partial y}\right)+\frac{\partial}{\partial z}\left(k_z\frac{\partial H}{\partial z}\right)+Q=m_w\gamma_w\frac{\partial H}{\partial t}, \quad (3)$$

where H is the total water head; k_x, k_y, and k_z are the permeability coefficients of the x, y, and z directions, respectively; t is time; Q is the source sink term, which is generally a function of position and time, and θ is the water content per unit volume.

3.3. Finite Element Analysis of the Three-Dimensional Seepage

In line with the variational principle, the three-dimensional seepage fixed solution problem is equivalent to solving the extreme value problem of the energy function when the coordinate axis direction is consistent with the principal seepage direction, namely:

$$I(H)=\iiint\limits_{\Omega}\frac{1}{2}\left[k_x\left(\frac{\partial H}{\partial x}\right)^2+k_y\left(\frac{\partial H}{\partial y}\right)^2+k_z\left(\frac{\partial H}{\partial z}\right)^2\right]dxdydz-\iint\limits_{\Gamma} qHds\Rightarrow \min, \quad (4)$$

where Γ is the boundary of the seepage zone.

In the process of groundwater seepage in the fractured rock mass, the seepage satisfies the continuity equation, the mass conservation equation, the initial conditions (Formula (5)), and the boundary conditions (Formula (6)).

$$H(x,y,z,t) = H_0(x,y,z,t_0)\ x,y,z \ni D \tag{5}$$

$$H(x,y,z,t) = \varphi(x,y,z,t)\ x,y,z \ni \Gamma \tag{6}$$

where D is the seepage zone, and φ is the known function of the water head boundary.

Using the finite element method to analyze the seepage field is the same as solving the extreme of the energy function, and the initial seepage field is inversed by the steady state analysis as the boundary head condition of the transient analysis. The seepage field in the engineering area is discretized, and then the subregions are superimposed. Assuming that the interpolation function is zero, the finite element equation for solving the seepage field is:

$$\text{Steady}-\text{state analysis}:\ [K]\{H\} = \{F\}, \tag{7}$$

and

$$\text{Transient analysis}:\ \Delta t([K]+[M])\{H_1\} = \Delta t\{Q_1\} + [M]\{H_0\}, \tag{8}$$

where $[K]$ is the overall penetration matrix, $\{H\}$ is the node head array, $\{F\}$ is the flow matrix, Δt is the time increment, $\{H_0\}$ is the water head at the beginning of the time increment, and $\{H_1\}$ is the water head at the end of the time increment.

In this study, the above equations were solved using finite element software, the natural seepage field was inverted by the steady state analysis equation, and the variations in the seepage fields for the different seepage control design schemes were analyzed with the transient analysis equation.

4. Three-Dimensional Transient Seepage of the Hydrogeological Models

4.1. Inversion Model

The distribution of the fracture network in the rock mass is chaotic. The question of how the fracture network should be reflected correctly in the calculation model is the most difficult problem when studying the seepage fields of fractured rock masses. Here, the inversion method was used to solve this problem. The permeability coefficient of each hydrogeological unit in the engineering area was deduced backward with the three-dimensional model from the conditions provided by the field geological exploration, borehole pressure water test, and the hydrogeological survey.

The boundary conditions of the model were determined using the hydrogeological survey data, and the borehole water pressure test value was used as the initial value of the permeability coefficient of the model. The regional natural seepage field was simulated, and ten representative water levels were selected and compared with the measured values of the water level of the seepage field. The permeability of the hydrogeological unit partition was adjusted, and the hydrogeological model that met the engineering requirements was established.

Structural planes, such as cross-connected or penetrating faults and fractures, may have roles as seepage channels in underground seepage. These discontinuous structural planes must be accurately reflected in the model in a practical engineering application before the calculation results of the model are used to guide engineering practice. To meet the scientific and engineering requirements of the numerical model, the model simulation was based on 1,003,384 discrete elements from the groundwater seepage system across the whole region (Figure 6) and included an area of 70 km^2. From the borehole water pressure test data and weathering zoning, the calculation area was divided into four layers in the vertical direction, and the bottom boundary was a slightly weathered rock mass, which was assumed to be an impervious boundary.

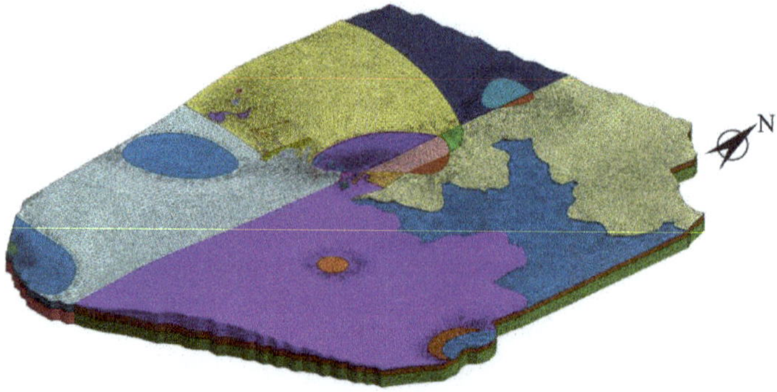

Figure 6. Three-dimensional hydrogeological model.

The permeability coefficient, k, is very important for seepage field analysis [32–34]. Based on the lithology distribution and the permeability coefficients of the engineering area, the model was divided into 6 parameter regions (Figure 7), in which the numbers 1–6 represent different hydrogeological partitions. Using the lithological characteristics of the engineering area, the field drilling water pressure test was carried out, and the permeability coefficient of each partition was calculated with the Babushkin formula. As the initial parameters of the model inversion, the inversion results were compared with ten monitoring points in the natural seepage field and the permeability coefficient of each partition was corrected (Table 2).

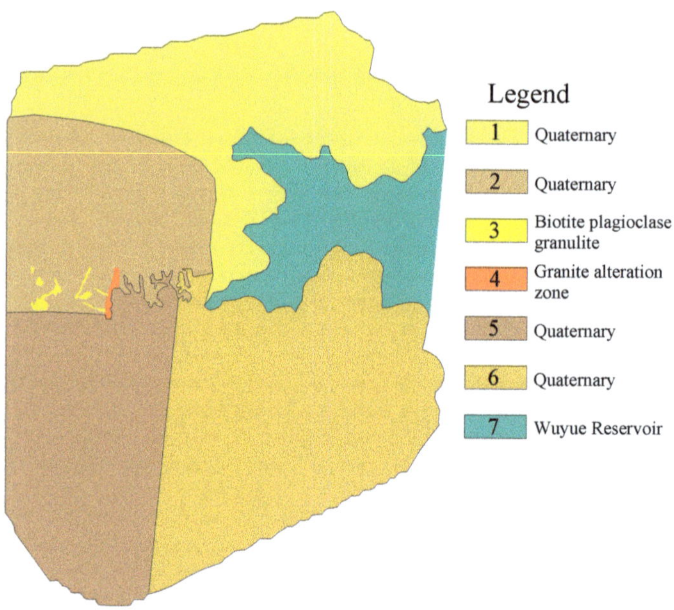

Figure 7. Hydrogeological model subdivision.

Table 2. Permeability coefficient of each layer.

Partition Number	The Permeability Coefficient of Principal Axis Direction (m/d)		
	k_{xx}	k_{yy}	k_{zz}
1	0.0232	0.0232	0.0232
2	2.592×10^{-7}	2.592×10^{-7}	2.592×10^{-7}
3	72.0576	72.0576	72.0576
4	8.64×10^{-6}	8.64×10^{-6}	8.64×10^{-6}
5	8.64×10^{-6}	8.64×10^{-6}	8.64×10^{-6}
6	0.0415	0.0415	0.0415

4.2. Validation of the Inversion Model

The natural seepage field and the simulated seepage field are shown in Figure 8. The numerical model was verified using ten identical monitoring points from Figure 8a,b that covered the water level values of features such as the engineering area, Wuyue Reservoir, and Laoshanzhai.

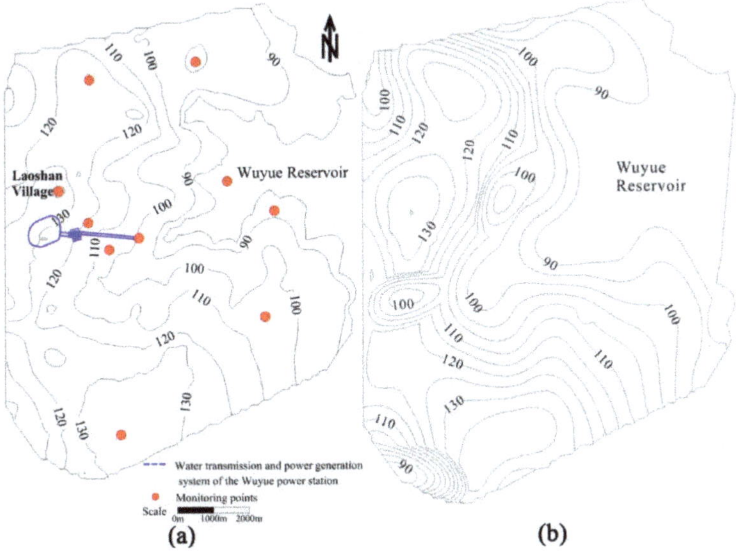

Figure 8. Comparison of (**a**) the natural seepage field and (**b**) the simulated seepage field.

The measured and calculated water levels of the monitoring points are shown in Figure 9a. The simulated water level was roughly consistent with the field-measured water level, and the average error of the simulated calculated water level was 1.98%.

Based on the rationality of the model parameter inversion, two seepage control schemes, namely normal seepage conditions and limited seepage conditions, were designed. The normal seepage conditions included five different transient flows, with transient flows of (1) 0–200 m^3/d, (2) 0–400 m^3/d, (3) 0–600 m^3/d, (4) 0–800 m^3/d, and (5) 0–1000 m^3/d. The limiting seepage conditions included three transient flows, with transient flows of (6) 0–2000 m^3/d, (7) 0–4000 m^3/d, and (8) 0–5000 m^3/d.

Three long observation holes, ZK01, ZK02, and ZK03, were set up on both sides of the tunnel axis in the project area to collect real-time, accurate information about the changes in the groundwater flow fields in the project area for a period of 2 years (Figure 4). Water level sensors (Figure 10) were installed to transmit the monitoring data in real-time and send out alarm information in time. The water levels of the natural flow fields of ZK01, ZK02, and ZK03 and the calculated water levels for the five seepage control design schemes are shown in Figure 10. This graph shows that the water level changes in the long observation holes fitted closely with seepage control design scheme 1.

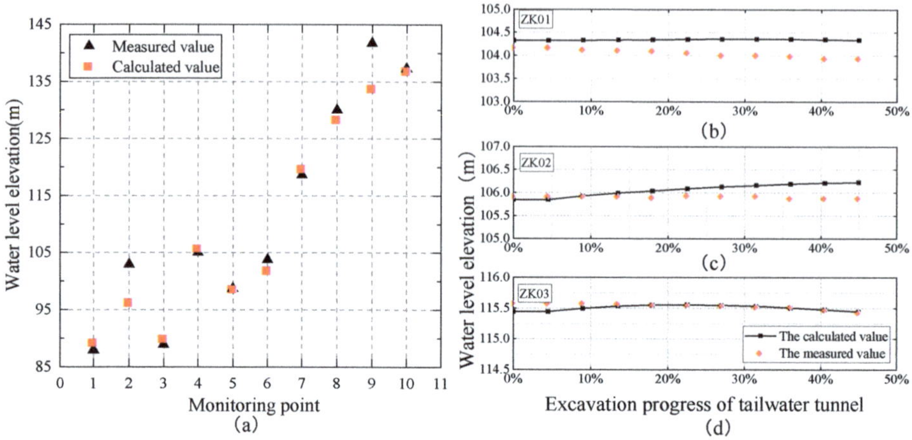

Figure 9. Comparison of the measured water level and the calculated water level. ((**a**) is comparison of measured and calculated groundwater flow field values, (**b**) is comparison of measured and calculated water level in long-term observation hole). (**c**) is comparison of measured and calculated water level in long—term observation hole ZK02, (**d**) is comparison of measured and calculated water level in long—term observation hole ZK03).

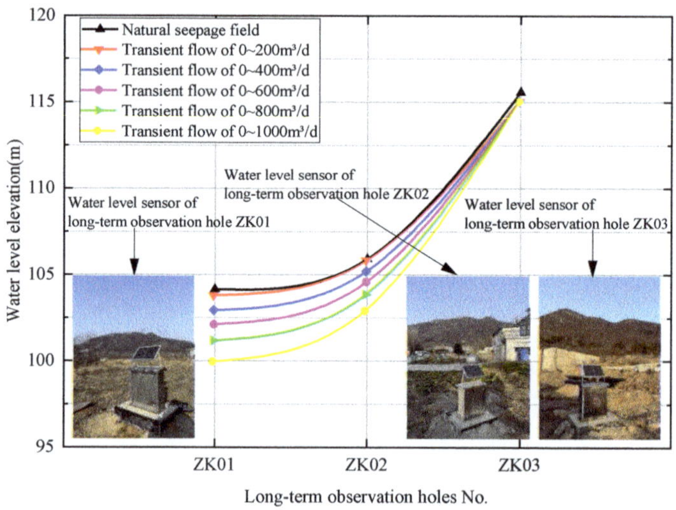

Figure 10. Water level changes in the long observation holes under different transient flows.

The variations in the water levels monitored in the long observation holes and calculated by the inversion model for seepage control design scheme 1 are shown in Figure 9b–d. Borehole ZK01 had a water level of 103.94 m and a simulated water level of 104.33 m, meaning a difference of 0.39 m. Borehole ZK02 had a water level of 105.88 m and a simulated water level of 106.22 m, meaning a difference of 0.34 m. Borehole ZK03 had a water level of 115.43 m and a simulated water level of 115.44 m, meaning a difference of 0.01 m. The average error between the simulated water level values and the sensor data was 0.16%, which shows that the model was satisfactory. The seepage parameters given by the inverted model were reasonable, and the values calculated for the three-dimensional seepage field model were reliable and suitable for use to simulate and predict the groundwater flow field in the engineering area. The numerical simulations can be used to quantify the variations in the seepage field and predict the water levels for a future phase of the tailwater tunnel construction.

4.3. Transient Seepage Analysis

Assuming that the inverted model produced results that were practical for the engineering, the hydrogeological model was used to predict the influence of normal seepage conditions on the changes in the groundwater seepage field during the excavation of the tailwater tunnel using a three-dimensional transient numerical simulation. The range of the influence of the tunnel excavation on groundwater was analyzed qualitatively, and the distance over which there were changes in the pore water pressure head because of the tunnel excavation was quantified and analyzed.

The depleted well points where the villagers were experiencing drinking water shortages were added to the model and simulated. Well points SJ96, SJ141, SJ146, and SJ202 were included in this analysis. Information on the distance between the well points and the axis of the tailwater tunnel is given in Table 3. For each of these wells, the patterns and changes in the water levels for the five seepage control schemes were analyzed.

Table 3. Drained civilian Wells.

Well Number	Location	Distance from Tunnel Axis (m)	Wellhead Elevation (m)	Depth of Well (m)
SJ96	Yangshuwan Village, south of the tunnel	750	113.0937	6.9
SJ97	Yangshuwan Village, south of the tunnel	790	110.3317	7.9
SJ141	Qiandawan village, south of the tunnel	101	114.6467	15
SJ142	Qiandawan village, south of the tunnel	105	115.8357	11
SJ146	Tulou Village, south of the tunnel	1150	111.4187	8.8
SJ202	Yangmingao village, north of the tunnel	1000	125.3427	7

The variations in the water level at well points SJ96, SJ141, SJ146, and SJ202 for the five seepage control schemes are shown in Figure 11. As shown in Figure 11a, the groundwater levels in well points SJ96, SJ146, and SJ202 were 0.054, 0.039, and 0.025 m, respectively. Figure 11 also shows that the groundwater levels were affected by the excavation of the tunnel, and the levels decreased as the distance from the tunnel site increased, and the rate of the decrease in the water level slowed as the distance from the tunnel increased. Figure 11a,c,e,g,i show that the water level decreased as the maximum transient flow increased. The water levels in well point SJ202 decreased by 0.025, 0.029, 0.031, 0.032, and 0.034 m for the five seepage control schemes.

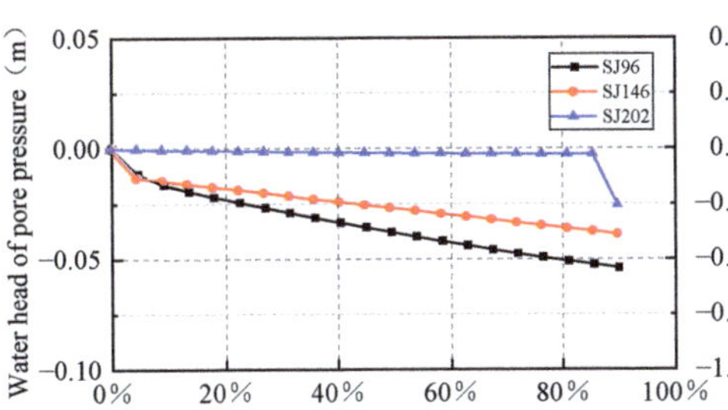

(a)

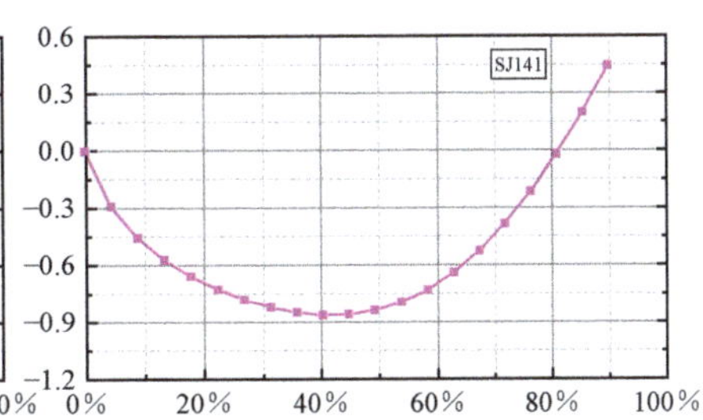

(b)

Figure 11. *Cont.*

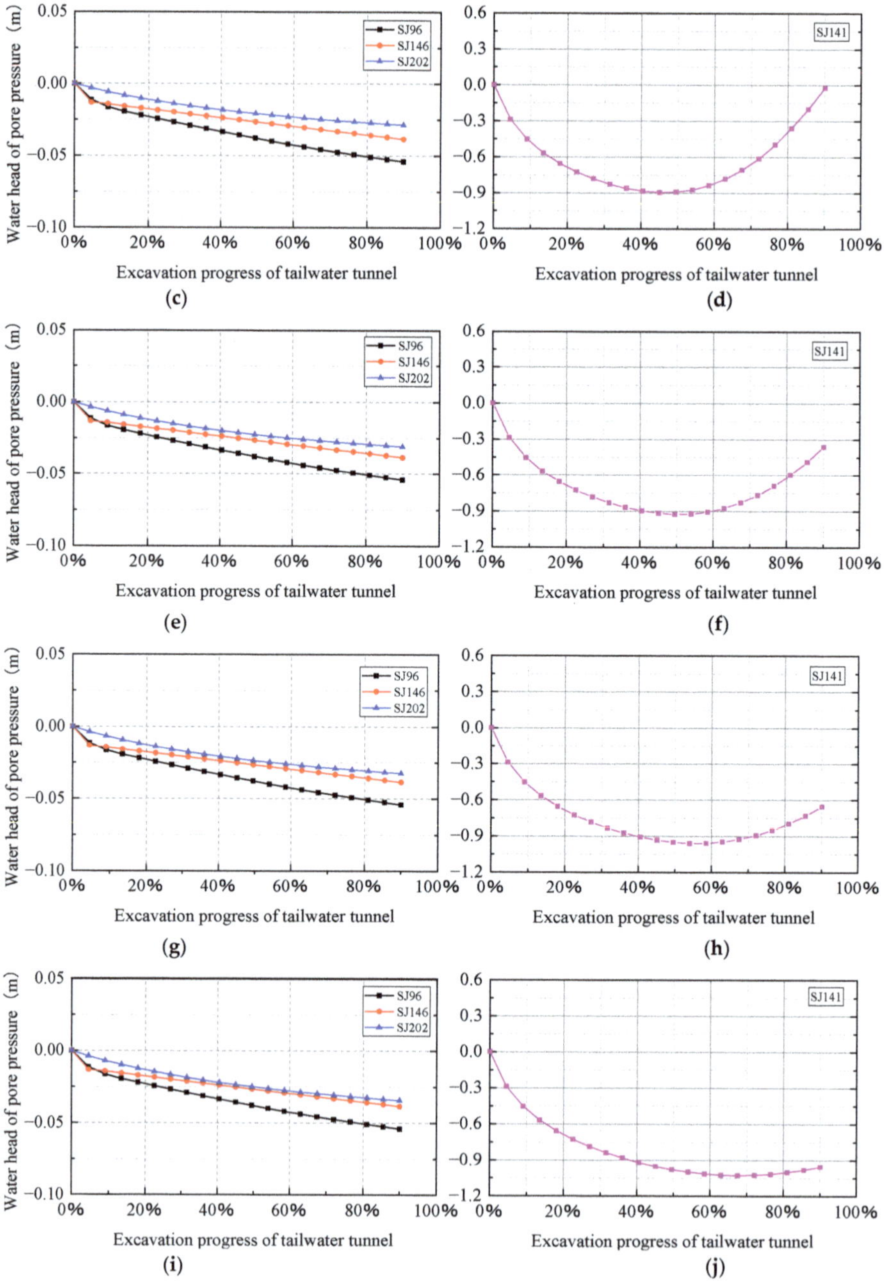

Figure 11. Changes in the water level for the different water inflow conditions. ((**a**) and (**b**) are changes in water level of the maximum water inflow of 200 m^3/d, (**c**) and (**d**) are the variation of water level for the maximum water inflow of 400 m^3/d, (**e**) and (**f**) are the changes in water level when the maximum inflow is 600 m^3/d, (**g**) and (**h**) are the changes of water level when the maximum water inflow is 800 m^3/d, (**i**) and (**j**) are the changes of water level when the maximum water inflow is 1000 m^3/d).

The water level in well point SJ141 first decreased and then increased as the tunnel excavation progressed, as shown in Figure 11b. Because of the excavation of the tunnel, a water inrush channel formed, the groundwater circulation system was disrupted, the groundwater seepage to the tunnel accumulated at a certain distance from the tunnel excavation, and the water level of the well point SJ141 decreased. As the distance from the tunnel excavation increased, the water inflow also increased, and the area of groundwater affected also increased. There was a large range in the distance of the groundwater seepage flows at places close to the tunnel (such as well point SJ141), resulting in temporary increases in the water level at well point SJ141. Across the whole process, the water level at well point SJ141 dropped by a maximum of 0.861 m.

As shown in Figure 11b,d,f,h,j, when the maximum water inflow of the tunnel increased from 200 m^3/d to 400, 600, 800, and 1000 m^3/d in turn, the water level in well point SJ141 first decreased and then gradually started to increase (Figure 11b) and then decreased slowly (Figure 11j). When the tunnel water inflow increased to a certain level, there was not enough groundwater in the immediate area to quickly supplement the head loss at well point SJ141, which then caused the water level of well point SJ141 to decrease slowly. The maximum water level decreases at well point SJ141 were 0.861, 0.894, 0.927, 0.959, and 1.028 m (Figure 11b,d,f,h,j).

The variations in the groundwater seepage field within 200 (S200), 300 (S300), and 400 m (S400) from the tunnel axis for the five seepage control schemes were simulated to provide qualitative information on how groundwater was impacted at different distances from the tunnel excavation. As shown in Figure 12, the excavation of the tunnel strongly influenced the groundwater seepage field at S200 and S300. When the maximum water inflow was 200 m^3/d, the water levels at S200 and S300 decreased by 0.070 and 0.024 m, respectively. As the water inflow increased, the decreases in the water levels were greater. When the water inflow reached 1000 m^3/d, the water levels at S200 and S300 decreased by 0.459 and 0.241 m, respectively. The groundwater level at S500 was only slightly affected by the excavation of the tunnel, as it was far from the tunnel axis. When the water inflow reached 1000 m^3/d, the water level only decreased by 0.061 m.

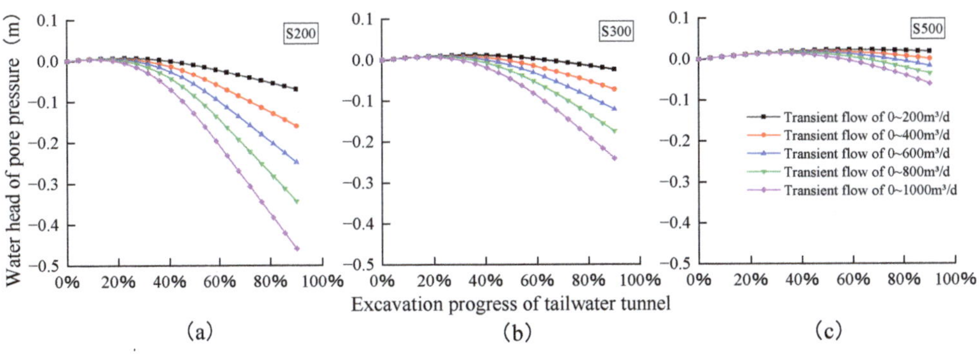

Figure 12. Variations in the water level at different distances under normal seepage conditions. ((**a**) is change of water level at 200m from the axis of the tunnel, (**b**) is change of water level at 300m from the axis of the tunnel, (**c**) is change of water level at 500m from the axis of the tunnel).

4.4. The Extent of the Influence on the Seepage Field

As the tunnel excavation progressed, the water inflow to the tunnel gradually increased. The maximum water inflow in the simulations was increased to 2000, 4000, and 5000 m^3/d, respectively, to gain information that could be used to help prevent accidents and avoid any risks from water inflow accidents during the tunnel construction. The distance to which the tunnel excavation affected the groundwater was qualitatively analyzed, the changes in the pore water pressure head caused by the tunnel excavation by distance were quantitatively assessed, and a formula was derived to quantify the distance of the influence.

The variations in the groundwater seepage field were simulated at 101 (SJ141), 200 (S200), 300 (S300), and 400 m (S400) from the tunnel axis under limiting seepage conditions. As shown in Figure 13, it can be seen that, for the three transient inflows, the water level at SJ141 fluctuated most and first increased and then decreased sharply. Because of the large water inflow, the groundwater flows to the tunnel affected a large area. As the water inflow channel continued to increase, the water inflow also continued to increase, thereby causing the water level at SJ141 to decrease sharply. For a maximum water inflow of 5000 m^3/d, the maximum water level at SJ141 was 3.842 m. There is a chance that engineering accidents may occur when a seepage channel forms and the water inflow increases; for example, the rock surrounding the tunnel could decrease, or the top of the tunnel could collapse. To account for these possible scenarios, scenarios for water inflows of more than 800 m^3/d were simulated and quantified.

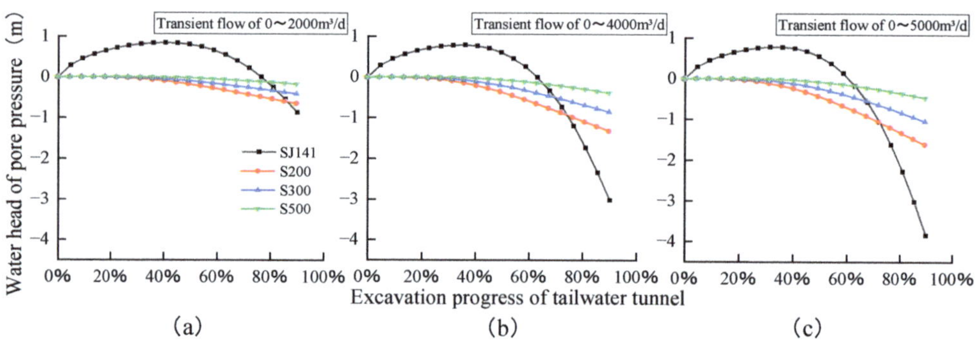

Figure 13. Variations in the water level at different distances under limiting seepage conditions. ((**a**) is change in water level of the maximum water inflow of 2000 m^3/d, (**b**) is change in water level of the maximum water inflow of 4000 m^3/d, (**c**) is change in water level of the maximum water inflow of 5000 m^3/d).

Given the potential influence of water inflows greater than 800 m^3/d on the project, the range of the influence of the tunnel excavation on groundwater was determined from the data from the inversion model, and a formula for calculating the range of the influence under different water inflows was derived. As shown in Figures 14 and 15, as the tunnel excavation progressed, the groundwater gradually decreased. When the maximum water inflow was 1000 m^3/d, the water level at a distance of 100 m from the tunnel axis decreased by 0.956 m. For a maximum water inflow of 5000 m^3/d, the water level at 200 m from the tunnel axis decreased by 1.604 m. Figures 14 and 15 also show that the pore pressure head gradually decreased until it stabilized as the distance from the axis of the tunnel increased, meaning that it was not affected by the tunnel excavation.

The formula for calculating the range of the influence of the tunnel excavation on the groundwater can be uniformly expressed as:

$$\mathrm{h} = \mathrm{a}\, x^2 + bx + c$$

where x is the distance from hole axis, m, and h is the pore pressure head, m. When h is 0, the pore pressure water is not affected by the tunnel excavation, and x is the linear distance to which the pore pressure water is affected by the tunnel excavation.

Regression analysis based on the least squares method was applied to the calculated data from the inverted model, and a formula was derived to show the range of the influence of the tunnel excavation on the groundwater, as shown in Table 4.

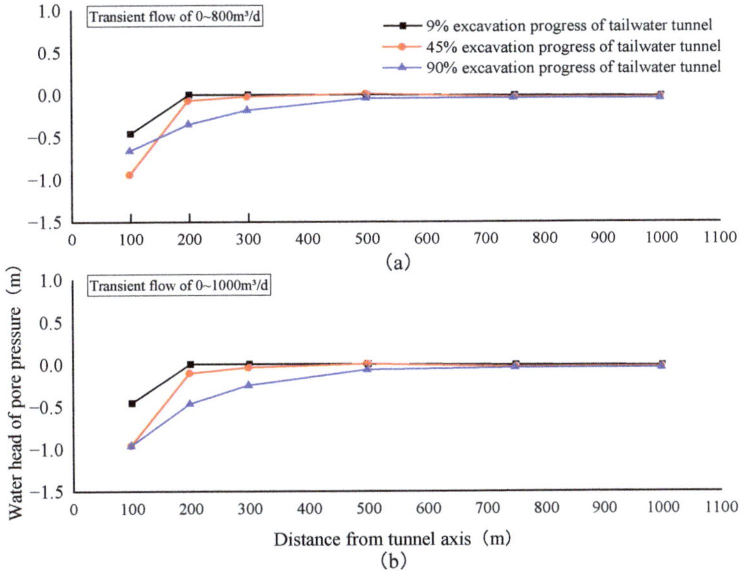

Figure 14. Variations in the water level under normal seepage conditions. ((**a**) is change in water level of the maximum water inflow of 800 m^3/d, (**b**) is change in water level of the maximum water inflow of 1000 m^3/d).

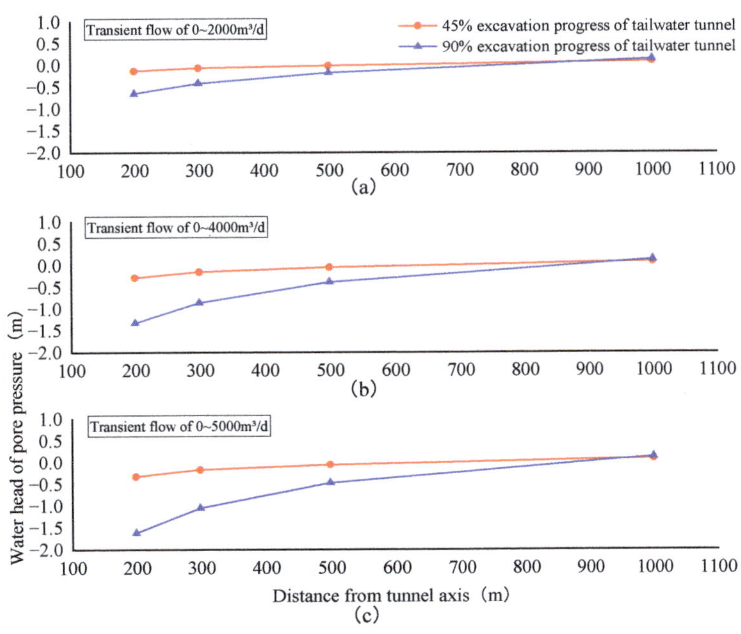

Figure 15. Variations in the water level under limiting seepage conditions. ((**a**) is change in water level of the maximum water inflow of 2000 m^3/d, (**b**) is change in water level of the maximum water inflow of 4000 m^3/d, (c) is change in water level of the maximum water inflow of 5000 m^3/d).

Table 4. Fitting formula of the range to which groundwater was influenced.

Maximum Inflow (m^3/d)	Fitting Formula of Groundwater Influence Area	R^2
800	$h = -1.618 \times 10^{-6}x^2 + 0.00234x - 0.798$	0.9338
1000	$h = -2.313 \times 10^{-6}x^2 + 0.00336x - 1.141$	0.9193
2000	$h = -1.228 \times 10^{-6}x^2 + 0.00243x - 1.060$	0.9948
4000	$h = -2.653 \times 10^{-6}x^2 + 0.00495x - 2.157$	0.9945
5000	$h = -3.305 \times 10^{-6}x^2 + 0.00608x - 2.643$	0.9943

5. Discussion

This study was initially designed to explore why drinking water had become depleted close to a construction site. The research area was defined by the extent of the initial hydrogeological survey. Field and laboratory experiments were carried out, and the results were combined to produce parameters such as the elastic modulus, Poisson's ratio, cohesion, the friction angle, and the permeability coefficient. A three-dimensional hydrogeological model was established from coordinates measured in three dimensions. A three-dimensional inversion method was used to verify the reliability of the model with natural seepage fields and high values for pore water. The distance to which groundwater was affected by increases in the water inflow in the tailwater tunnel was then predicted. Based on the patterns of change in the simulated water level, the least squares method was used for data fitting, and a formula was proposed for calculating the distance and extent to which groundwater was influenced for a range of water inflow scenarios.

Groundwater flow field: the average error of simulated water level is 1.98%; Long-term monitoring of water level: the average error of simulated water level is 0.16%. In this study, the impact of the tailwater tunnel excavation on groundwater was quantified. This represents the first step of the research. Further studies will explore the feasibility of using different methods for excavating the tunnel, and results from time–frequency domain analysis of the vibration effects and vibration energy will be combined to improve the accuracy of current knowledge about the degree and distance to which tunnel excavation might impact human settlements. These studies should contribute to achieving the goal of cleaner production, thereby achieving the goal of peak carbon and carbon neutrality.

6. Conclusions

In this study, a regional hydrogeological model was established by inverting the parameters of the natural flow field of a region. The model was then used to numerically simulate the distribution and variations in seepage fields that formed during the tunnel excavation under normal seepage conditions and limiting seepage conditions. The model was validated, and error analysis showed that the approach was suitable. The results for the two seepage conditions were compared to understand the range of the influence on the groundwater flow, and a formula was proposed to calculate the distance and degree to which groundwater was affected by the tunnel excavation.

The main conclusions of this study are as follows.

(1) The project is located in a mountainous area with complex topography, so it is impossible to analyze or assess the geology and hydrogeology of the whole study area. Here, from a field investigation, in situ tests, laboratory tests, and numerical simulations, an inverted model was established and verified, and simulations were produced to facilitate a study of how the tunnel excavation influenced groundwater in the area. This novel approach could be used for further studies of seepage fields in hydraulic tunnels.

(2) To understand the depletion of domestic wells and to determine what caused the seepage in the regional flow field, the changes in the water level at different well points were quantified. The results showed that after the tailwater tunnel was excavated, the groundwater levels in well points SJ96, SJ146, and SJ202 dropped to 0.054, 0.039, and 0.025 m, respectively.

(3) The distance to which the tunnel construction affected the groundwater was then quantified. From these results, the variations in the groundwater flow field in the engineering area for different water inflow scenarios were simulated. The results showed that, for water inflows of 200, 400, 600, 800, and 1000 m^3/d, the water level of SJ141 dropped by 0.861, 0.894, 0.927, 0.959, and 1.028 m, respectively, and the water level of well point SJ202 decreased by 0.025, 0.029, 0.031, 0.032, and 0.034 m, respectively.

(4) The influence of limiting seepage conditions on the seepage field at different distances in the engineering area was then analyzed, and a formula was derived for calculating the distance to which groundwater was affected for a range of different water inflows. A 'smart site' was established using the simulation outputs and real-time water level data from the three observation holes, ZK01, ZK02, and ZK03, to provide visual information and support the project management.

(5) In following studies, data for the vibration effects of tunnel drilling and blasting will be combined with time–frequency domain analysis to obtain improved information about the degree and distance to which the tunnel excavation affects human settlements. Further studies will also contribute to achieving a goal of cleaner production, which will then help to realize the goal of peak carbon and carbon neutrality.

Author Contributions: Conceptualization, T.J.; Data curation, X.P.; Funding acquisition, T.J.; Investigation, X.P., W.W., L.L., and S.G.; Methodology, T.J. and X.P.; Project administration, T.J.; Software, X.P.; Supervision, T.J.; Validation, X.P.; Writing—original draft, X.P.; Writing—review and editing, X.P. All authors have read and agreed to the published version of the manuscript. All authors have read and agreed to the published version of the manuscript.

Funding: This research was funded by the National Natural Science Foundation of China (Grant number 42090052) and the Henan Institution of Higher Education Key Scientific Research Project (22A170013). And The APC was funded by the National Natural Science Foundation of China.

Institutional Review Board Statement: Not applicable.

Informed Consent Statement: Not applicable.

Data Availability Statement: The data used to support the findings of this study are included within the article.

Acknowledgments: In the field investigation, Xiangjing Lu provided much assistance in the fieldwork, which positively influenced the safety and physical and mental health of the investigators. We would like to express our heartfelt thanks to Xiangjing Lu for his assistance in this work.

Conflicts of Interest: The authors declare that they have no conflict of interest.

References

1. Tang, Y.G.; Zhang, Q.; Qi, J.H.; Xu, M.; Li, X.; Qu, C.H.; Yi, L.; Wang, D. Change Analysis of Karst Landforms, Hydrogeological Conditions and Effects of Tunnel Excavation on Groundwater Environment in Three Topography Grades in China. *Water* **2023**, *15*, 207. [CrossRef]
2. Liu, J.Q.; Sun, Y.K.; Li, C.J.; Yuan, H.L.; Chen, W.Z.; Liu, X.Y.; Zhou, X.S. Field monitoring and numerical analysis of tunnel water inrush and the environmental changes. *Tunn. Undergr. Space Technol.* **2022**, *122*, 104360. [CrossRef]
3. Zheng, Z.; Liu, R.; Li, S.; Zhang, Q. Tunnel face failure mechanism with sand layer partial collapse. *Arab. J. Geosci.* **2020**, *13*, 1077. [CrossRef]
4. Yu, J.; Ge, S.; Hou, L.; Zhang, Y.; Liu, Y. Limit Drainage of Separated Tunnel Based on Ecological Groundwater Table: A Case Study. *Adv. Civ. Eng.* **2021**, *2021*, 2669924. [CrossRef]
5. Liu, W.; Zhu, J.; Zhang, H.; Ma, X.; Xie, J. Geological conditions of saturated soft loess stratum and influence of tunnel excavation and dewatering system on its groundwater environment. *Bull. Eng. Geol. Environ.* **2022**, *81*, 128. [CrossRef]

6. Wu, J.; Lu, Y.; Wu, L.; Han, Y.; Sun, M. Numerical Investigation of Water Inflow Characteristics in a Deep-Buried Tunnel Crossing Two Overlapped Intersecting Faults. *Water* **2023**, *15*, 479. [CrossRef]
7. Fang, H.; Zhu, J.T.; Qiu, J.R.; Su, Y.M.; Xia, D. Laminar and turbulent groundwater flows in confined two- and three-dimensional discrete fracture networks. *Eng. Geol.* **2022**, *308*, 106824. [CrossRef]
8. Zhang, Z.; Zhang, M.; Li, Z.; Zhao, Q.; Pan, Y.; Wang, J. Distribution for hydraulic head on tunnel structures in water-rich mountainous region considering influences of fault geology using virtual image technique. *Environ. Earth Sci.* **2021**, *80*, 528. [CrossRef]
9. Huang, D.; Huang, W.-B.; Ke, C.-Y.; Song, Y.-X. Experimental investigation on seepage erosion of the soil–rock interface. *Bull. Eng. Geol. Environ.* **2021**, *80*, 3115–3137. [CrossRef]
10. Wang, Y.; Li, Z.; Jing, H.; Li, Y.; Wang, M. Study on the seepage characteristics of deep buried tunnels under variable high-pressure water heads. *Bull. Eng. Geol. Environ.* **2020**, *80*, 1477–1487. [CrossRef]
11. Liu, J.; Liu, D.; Song, K. Evaluation of the Influence Caused by Tunnel Construction on Groundwater Environment: A Case Study of Tongluoshan Tunnel, China. *Adv. Mater. Sci. Eng.* **2015**, *2015*, 149265. [CrossRef]
12. Chen, L.; Wang, Z.-F.; Wang, Y.; Bai, X.; Lai, J. Characteristics and failure analysis of a railway tunnel collapse influenced by cavity in phyllite strata. *Eng. Fail. Anal.* **2022**, *142*, 106794. [CrossRef]
13. Intui, S.; Inazumi, S.; Soralump, S. Evaluation of Land Subsidence during Groundwater Recovery. *Appl. Sci.* **2022**, *12*, 7904. [CrossRef]
14. Chen, N.-C.; Wen, H.-Y.; Li, F.-M.; Hsu, S.-M.; Ke, C.-C.; Lin, Y.-T.; Huang, C.-C. Investigation and Estimation of Ground-water Level Fluctuation Potential: A Case Study in the Pei-Kang River Basin and Chou-Shui River Basin of the Taiwan Mountainous Region. *Appl. Sci.* **2022**, *12*, 7060. [CrossRef]
15. Liu, J.-Q.; Liu, C.; Liu, X.-Y.; Wang, S.; Yuan, H.-L.; Li, C.-J.; Dong, J.-L. Prediction of water–mud inrush hazard from weathered granite tunnel by an improved seepage erosion model. *Bull. Eng. Geol. Environ.* **2021**, *80*, 9249–9266. [CrossRef]
16. Wang, S.; Zhang, Q.; Zhao, L.; Jin, Y.; Qian, J. Seepage Characteristics Study of Single Rough Fracture Based on Numerical Simulation. *Appl. Sci.* **2022**, *12*, 7328. [CrossRef]
17. Preisig, G.; Dematteis, A.; Torri, R.; Monin, N.; Milnes, E.; Perrochet, P. Modelling Discharge Rates and Ground Settlement Induced by Tunnel Excavation. *Rock Mech. Rock Eng.* **2014**, *47*, 869–884. [CrossRef]
18. Heuer, R.E. Estimating rock tunnel water inflow. *Rapid Excav. Tunn. Conf.* **1995**, *41–60*, 10459065.
19. Raymer, J. Groundwater inflow into hard rock tunnels. *Tunn. Tunn. Int.* **2001**, *33*, 50–53.
20. Wu, Z.; Fang, F.; Li, X.; Xiao, H.; Liu, X.; Rao, Y.; Li, Y.; Wang, J.; Luo, Y.; Li, Z. Division Method and Seepage Law of Seepage Channels in a Tight Reservoir. *Geofluids* **2015**, *2021*, 4804513. [CrossRef]
21. Dong, T.; Cao, P.; Gui, R.; Lin, Q.; Liu, Z. Experimental Study on Permeability Coefficient in Layered Fine Tailings under Seepage Condition. *Geofluids* **2021**, *2021*, 8850138. [CrossRef]
22. Wang, W.; Xue, J.; Qingyang, H.; Jiang, T.; Dong, J. Dynamic and Static Reserve Recharge Characteristics of a Dewatering Well Seepage from an Aquifer Bottom by Sand Tank Seepage Experiment. *Geofluids* **2022**, *2022*, 3971570. [CrossRef]
23. Qiu, J.; Zheng, D.; Zhu, K. Seepage Monitoring Models Study of Earth-Rock Dams Influenced by Rainstorms. *Math. Probl. Eng.* **2016**, *2016*, 1656738. [CrossRef]
24. Ma, H.F.; Yao, F.F.; Niu, X.G.; Guo, J. Experimental Study on Seepage Characteristics of Fractured Rock Mass under Different Stress Conditions. *Geofluids* **2021**, *2021*, 6381549.
25. Chae, G.-T.; Yun, S.-T.; Choi, B.-Y.; Yu, S.-Y.; Jo, H.-Y.; Mayer, B.; Kim, Y.-J.; Lee, J.-Y. Hydrochemistry of urban groundwater, Seoul, Korea: The impact of subway tunnels on groundwater quality. *J. Contam. Hydrol.* **2008**, *101*, 42–52. [CrossRef] [PubMed]
26. Gisbert, J.; Vallejos, A.; González, A.; Bosch, A.P. Environmental and hydrogeological problems in karstic terrains crossed by tunnels: A case study. *Environ. Geol.* **2009**, *58*, 347–357. [CrossRef]
27. Vincenzi, V.; Gargini, A.; Goldscheider, N. Using tracer tests and hydrological observations to evaluate effects of tunnel drainage on groundwater and surface waters in the Northern Apennines (Italy). *Hydrogeol. J.* **2009**, *17*, 135–150. [CrossRef]
28. Li, H.; Kagami, H. Groundwater level and chemistry changes resulting from tunnel construction near Matsumoto City, Japan. *Environ. Geol.* **1997**, *31*, 76–84.
29. Bai, Y.; Wu, Z.; Huang, T.; Peng, D. A Dynamic Modeling Approach to Predict Water Inflow during Karst Tunnel Excavation. *Water* **2022**, *14*, 2380. [CrossRef]
30. Zheng, X.; Yang, Z.; Wang, S.; Chen, Y.-F.; Hu, R.; Zhao, X.-J.; Wu, X.-L.; Yang, X.-L. Evaluation of hydrogeological impact of tunnel engineering in a karst aquifer by coupled discrete-continuum numerical simulations. *J. Hydrol.* **2021**, *597*, 125765. [CrossRef]
31. Huang, Z.; Zhao, K.; Li, X.Z.; Zhong, W.; Wu, Y. Numerical characterization of groundwater flow and frac-ture-induced water inrush in tunnels. *Tunn. Undergr. Space Technol.* **2021**, *116*, 104119. [CrossRef]
32. Dou, H.-Q.; Han, T.-C.; Gong, X.-N.; Qiu, Z.-Y.; Li, Z.-N. Effects of the spatial variability of permeability on rainfall-induced landslides. *Eng. Geol.* **2015**, *192*, 92–100. [CrossRef]

33. Yang, T.; Liu, H.; Tang, C. Scale effect in macroscopic permeability of jointed rock mass using a coupled stress–damage–flow method. *Eng. Geol.* **2017**, *228*, 121–136. [CrossRef]
34. Tang, M.; Xu, Q.; Yang, H.; Li, S.; Iqbal, J.; Fu, X.; Huang, X.; Cheng, W. Activity law and hydraulics mechanism of landslides with different sliding surface and permeability in the Three Gorges Reservoir Area, China. *Eng. Geol.* **2019**, *260*, 105212. [CrossRef]

applied sciences

MDPI

Communication

Capillary Rise in Layered Soils

Zhenhua Zhao [1,2], Zhenjiang Luo [1,2,*], Hongjie Sun [1,2], Haitao Li [1,2], Qiang Liu [3] and Haiyan Liu [3]

[1] 801 Institute of Hydrogeology and Engineering Geology, Jinan 250000, China
[2] Shandong Engineering Research Center for Environmental Protection and Remediation on Groundwater, Jinan 250000, China
[3] College of Earth Science and Engineering, Shandong University of Science and Technology, Jinan 250000, China
* Correspondence: luozhenjiang89@gmail.com

Abstract: Capillary rise tests were conducted on soil columns containing of three layers of sandy soils with coarser over finer over coarser sandy soil to investigate the effect of the relatively finer soil interlayer. The capillary rise height, rate, and water distribution were observed in laboratory tests of four layered soil columns, with two homogeneous (without the interlayer) soil columns serving as the controls. The final maximum height of the capillary rise in the soil column with the interlayer was larger than that of the column without the interlayer when the interlayer was laid around the water entry value of the underlying soil. The water content was not continuous in the entire soil profile with the interlayer, and a small matric suction gap was observed in the relatively fine soil between the soil column with and without the interlayer.

Keywords: unsaturated soil; capillary rise; layered soil; water entry value

check for updates

Citation: Zhao, Z.; Luo, Z.; Sun, H.; Li, H.; Liu, Q.; Liu, H. Capillary Rise in Layered Soils. *Appl. Sci.* **2023**, *13*, 3374. https://doi.org/10.3390/app13063374

Academic Editors: Bing Bai, Daniel Dias, Angeles Sanroman Braga and Tiago Miranda

Received: 10 August 2022
Revised: 7 January 2023
Accepted: 16 January 2023
Published: 7 March 2023

1. Introduction

The capillary rise of water is an important phenomenon, as it can enhance the soil freezing, thawing, and settlements of buildings. It can also cause damage to both concrete and steel if the water contains erosive ions, or influences the water balance and management in irrigated areas with shallow groundwater tables [1] To evaluate the water capillarity of the soil, the maximum capillary height, the water content profile, and the rate of capillary rise are the three keys. Hird and Bolton [2] used the column test, which controlled the relative air humidity above sample surface and the temperature of the entire sample, to discuss the capillary height. Polansky and Kaya [3] proposed a model to predict the overall behavior of the capillary rise dynamics in a heated capillary tube. Liu et al. [4] proposed a new theoretical solution for quick and easy estimation of the maximum height of the capillary rise in homogeneous soil. However, the soil deposit is usually layered. Soil texture has a decisive effect on the hydraulic properties of soil, such as unsaturated hydraulic conductivities ($k(h)$), soil water retention curve (SWRC), field capacity, air entry value, and capillary rise height and rate [5]. Typically, the finer the soil texture is, the greater the porosity, the water retention capacity, and the maximum capillary height, but the permeability and the capillary rise rate are slower. the capillary rise height can also be influenced by the concentration of iron in the water [6,7] and the thermal conductivity [8].

Layering in the soil can affect both the downward infiltration and the upward capillary rise. Vertical infiltration in layered soils has been studied by many researchers [9–11]. All of these studies have reported that the infiltration in layered soils is complex because of the high nonlinearity of SWRC, hydraulic conductivity, various boundary and initial conditions, the textural contrast, and the hysteresis of SWRC, which increases the complexity further [12]. Compared with extensive research on the infiltration, the related reports about capillary rise in layered soils are relatively less, particularly on the water capillary rise process without evaporation. In a study of the capillary height in layered soil, the steady state capillary rise was strongly influenced by the depth of the water table, and soil with a finer layer overlying a coarser layer, a linear relationship was observed between the thickness of the coarser layer and the critical groundwater depth, below which capillary water could not enter into the upper finer layer [13]. In a study of the water content profile, Shokri et al. [14] demonstrated that a capillary pressure jump caused by the air invading the interface between the finer and the coarser layers, and that will significantly modify the water distribution as compared to the homogeneous soil profile.

In the present study, laboratory column tests and numerical simulations were used to investigate the characteristics of the water capillary rise in layered soils. The results show that the soil suctions on the two sides of the interface, where is higher than the maximum capillary rise height of the coarser soil, are not same when water flow reaches the hydrostatic equilibrium condition.

2. Experiments

2.1. Material Properties

In this study, two types of commercially available sandy soil, called k-7 and k-8, were used. The grain size distributions for the two soils are shown in Figure 1. According to the standard (Japanese Geotechnical Society 0131 [15]), the k-7 soil is a sand (S-F) with a fine fraction (<0.075 mm) of around 10%, and the k-8 soil is a fine sand (SF), if British standard (EN ISO 14688-2 [16]) is applied, the k-7 soil can be classified as even-graded sand, and k-8 soil is a medium-graded sandy silt. The k-7 soil, and the k-8 soil have specific gravities of 2.67 and 2.69, respectively. Saturated hydraulic conductivity tests [17] were conducted on the k-7 soil by using the constant-head method and on the k-8 soil with the falling-head method. In constant-head test setup, the water supply at the inlet was adjusted to make the difference of head between the inlet and outlet remains constant during the test procedure. While in falling-head test setup, water from a standpipe flows through the soil and the water head varies with time. The saturated hydraulic conductivity of both soils was determined using the arithmetic mean of three tests at a dry density of 1.58 g/cm^3. The k-7 soil had a saturated hydraulic conductivity of 1.5×10^{-3} cm/s, and the value of the k-8 soil was 2.2×10^{-5} cm/s.

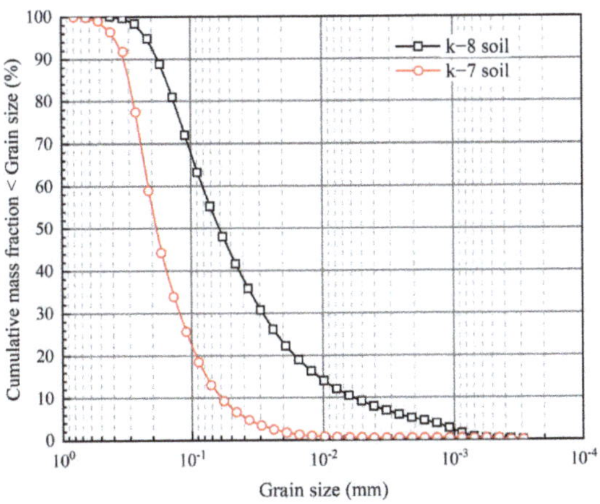

Figure 1. Grain size distributions for the soils used in the column capillary rise tests.

The soil water retention data of the soils were obtained during the wetting (sorption) process by using the two homogeneous soil columns, and then, the data points were fit to the van Genuchten model [18] by using the retention curve software (RETC [19]) to obtain the soil water retention curves (Figure 2a). The volumetric water content, θ, and the suction head, h, were related as follows:

$$\theta = \theta_r + \frac{(\theta_s - \theta_r)}{\left[1 + (\alpha h)^n\right]^m} \quad (1)$$

where θ_s and θ_r indicate the saturated and the residual values of the volumetric water content, respectively. α, m, and n are the fitting parameters. The hydrologic parameters are given in Table 1. By adopting the hydraulic conductivity model proposed by Mualem [20], we predicted the hydraulic conductivity of a soil as a function of either the volumetric water content or the suction head by using these fitting parameters and the saturated hydraulic conductivity, k_{sat}. When the pore-connectivity parameter, l, was approximately 0.5 for many soils as originally proposed by Mualem [20], in terms of the suction head, the unsaturated hydraulic conductivity could be expressed as follows:

$$k(h) = k_{sat}\frac{\left\{1 - (\alpha h)^{nm}\left[1 + (\alpha h)^n\right]^{-m}\right\}^2}{\left[1 + (\alpha h)^n\right]^{\frac{m}{2}}} \quad (2)$$

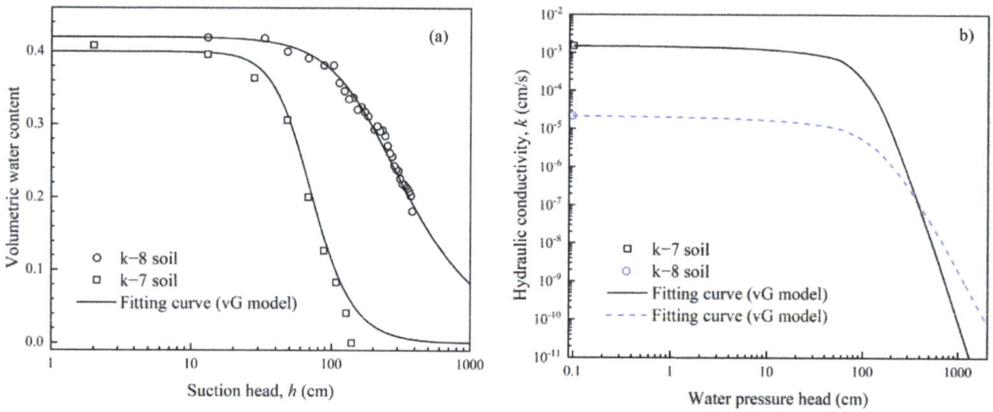

Figure 2. (**a**) Soil water retention curves (main wetting), and (**b**) unsaturated hydraulic conductivity curves for the k-7 and k-8 soils.

Table 1. Parameters applicable to the main wetting curve.

Soil	θ_s	θ_r	α (cm^{-1})	n	l
k-7	0.408	0	0.016	3.420	2.1
k-8	0.420	0.001	0.005	1.929	1

This equation cannot be solved unless the value of m − 1 + 1/n is an integer. The simplest case is for the value of 0, which leads to m = 1 − 1/n. Under this limit, the unsaturated hydraulic conductivities as a function of the suction head following a wetting path are as given in Figure 2b.

The soil water retention data were measured by gravimetric sampling (Figure 2a). As shown in the Figure 2a, under the same suction, the k-8 soil has a larger volumetric water content than that of the k-7 soil, which means thar the k-8 soil has the higher water retention capacity. The air entry value of the k-7 soil is about 30 kPa, for the k-8 soil, the value is about 100 kPa, at same time under this suction, the k-7 soil is nearly residual condition. As shown in the Figure 2b, the saturated hydraulic conductivity of the k-7 soil is 100 times over that of the k-8 soil; however, the value decays quickly. Conversely, the hydraulic conductivity of the k-8 soil decrease slowly, therefore, the two curves intersect at the point where the suction is around 100 kPa.

2.2. Capillary Rise Experiments

The capillary rise experiments were conducted in the plexiglass column (Figure 3). Each column was composed with the same unit; the inner diameter and height were 100 mm and 200 mm, respectively, except at the position of the holes, where the time domain reflectometry (TDR) sensors were inserted. A rubber O-ring was placed in the groove, and the column units were connected to each other by bolts. The soils were first oven dried at 105 °C for 24 h (2017) and then placed into the columns. The soils were densified by vibration with a rubber hammer to achieve the target dry density of 1.58 g/cm^3 for both the k-7 soil and the k-8 soil. As shown in Figure 4, four layered soil cases were set up. A porous stone was placed at the bottom of the columns, and a lip with a 5-mm-diameter hole was fixed on the top of the columns. A constant 2-cm positive water head was maintained at the bottom of the columns during the entire test procedure. Each layered case adopted the stratified structure as a sandwich with a 20-cm-thick middle layer of the k-8 soil and a layer of the k-7 soil on the other two sides. The difference among these cases was the height where the k-8 soil was laid; the k-8 soil layer was laid from 20 cm, 60 cm, 100 cm, and 120 cm in columns A, B, C, and D, respectively. Twelve water content sensors (TDR (EC-5, Decagon Co., Ltd., Pullman, WA, USA)) were installed in each column. Two other homogeneous soil columns were set up as the controls by using the same preparation procedure as that of the other four columns. The height of the homogeneous k-7 soil column was 2 m and that for the homogeneous k-8 soil column was 3 m at first. However, the capillary water reached the top surface of the soil profile. To avoid the effect of evaporation, the homogeneous k-8 column was tested for the second time, and the height was 4 m. The data were automatically obtained by the data logger every 5 min. After 240 days, the final water contents were directly measured by disassembling the column and oven method along the center of the columns at the height where the TDR sensors were installed.

Figure 3. Setup of the capillary rise in the layered soil.

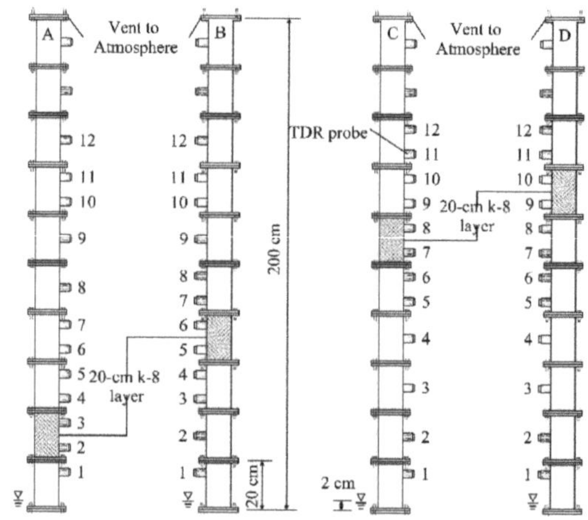

Figure 4. Schematic of the setup of the layered soil columns.

3. Numerical Simulation

The software Hydrus (2D/3D) [21] was used in this study. Based on the Richards equation [22], the numerical model was established following the experimental work, as described above. The upper boundary was set as the atmospheric condition without evaporation. The lower boundary was assigned to the condition of constant saturated water content. The entire finite element method (FEM) mesh was discretized uniformly except for the interface of the two soils, where a mesh refinement was added. The initial volumetric water content of the FEM column was set to 0.003, which was a little larger than the residual water content of both the soils. Next, 12 observation points were added to each FEM unit, at different elevations that were the same as the location of the TDR sensors in the physical model.

4. Results and Discussions

In all of the columns, the capillary water increased gradually with time (Figure 5a). The capillary water entered the finer interlayer (k-8 soil in this case) if it was laid lower than the maximum capillary height of the underlying coarser soil. The highest maximum capillary height, 153 cm, which was higher than the maximum capillary height of the homogeneous k-7 soil, was obtained in column D. However, in the other three columns, the maximum capillary height was the same as that of the homogeneous k-7 soil column. Therefore, once the water flowed into the finer soil, whether the maximum capillary height of layered soil was higher than that of the homogeneous soil depended on the height where the finer layer was laid. If the sum of the finer layer thickness and the coarser layer thickness (distance from the water level) was less than the maximum capillary height of the coarser layer, the maximum capillary height of layered soil was equal to that of the homogeneous coarser soil; otherwise, the former was larger.

The rate of capillary rise was different and not constant for each case. The capillary rate showed a power function relationship to the elapsed time; the corresponding regression equation is shown in Figure 5b. The tendency of the capillary rise rate in all of the layered cases was similar, which might be attributed to the fact that the thickness of the finer layer (20 cm) occupied a small portion of the total layered soil height, except in the case of column A. For column A, the finer soil layer was laid at a low position near the water level; it limited the water flow velocity because of its own lower hydraulic conductivity compared with that of the coarser soil in the same suction value.

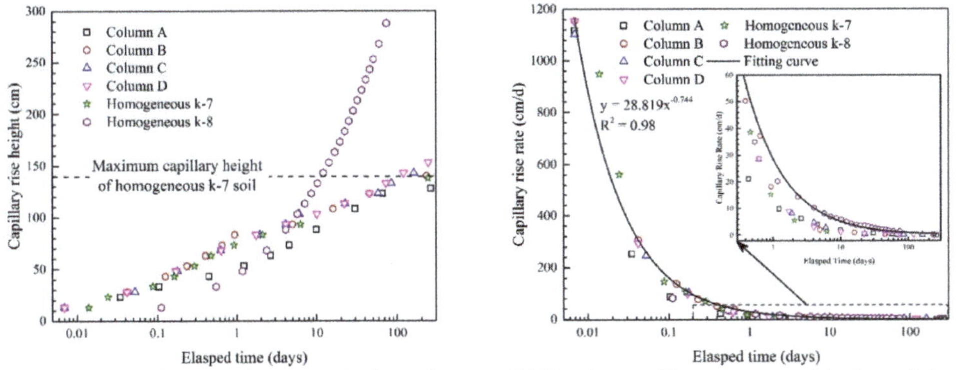

(**a**) Varying capillary rise height with elapsed time (**b**) Varying capillary rise rate with elapsed time

Figure 5. Comparison of the capillary rise height (**a**) and the rate (**b**) between the layered soil profile and the homogeneous soil profile.

Figure 6 shows the volumetric water content profiles of all of the six columns. The two solid curves are the water content profiles of the homogeneous k-7 and k-8 soil column, and the results of the other four layered cases are drawn as scatter plots. The water content value of the k-7 soil at the height greater than the maximum capillary height of the homogeneous k-7 soil was approximately equal to the residual water content. The water could be stored in the finer interlayer, and the amount of water was larger than that of the k-7 soil at the same height, but lower than that of the homogeneous k-8 soil at the same height. This result indicated a "gap" of the water content or suction between the interlayer k-8 soil and the homogeneous k-8 soil. According to the common view, the suction at the interface should be equal at two sides, therefore the water content at the two sides should be equal to the value of the homogeneous soil column at the height when the water flows reach at hydrostatic equilibrium condition. In other words, based on the observed water content, the suction of the layered soil might not be continuous because of the interface between the two types of soils due to the very different between the air entry values of the two kinds of soils.

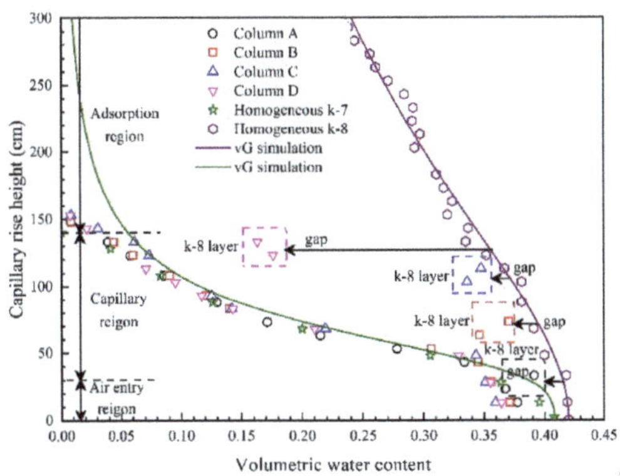

Figure 6. Water content distribution in the layered soil profile after the water capillary rise for 240 days.

As shown in Figure 7, generally, at most of the observed points, the numerical results were consistent with the test results. However, the simulated results of the finer soil layer (k-8 layer) were lower than that of the test results (Figure 7b). This implied that even if the parameters in the simulation were same, it was difficult to ensure that the simulated results were consistent in all of the test cases. The effect of the interface played an important role in determining the characteristics of the water flow. Figure 7b shows that the measured volumetric water content increased with elapsed time. However, at observation point No. 4 (53 cm above the water level), the curve was not like the other observed results, which showed smooth increases; there was an obvious "slow down" process. The process started just before the water arrived at the adjacent higher observation point (No. 5, 63 cm above the water level) and ended when the volumetric water content of observation point No. 5 became steady. This result revealed that when the wetting front went through the interface between the coarser and the finer soil layer, the limit water from the lower section gave priority to the finer soil.

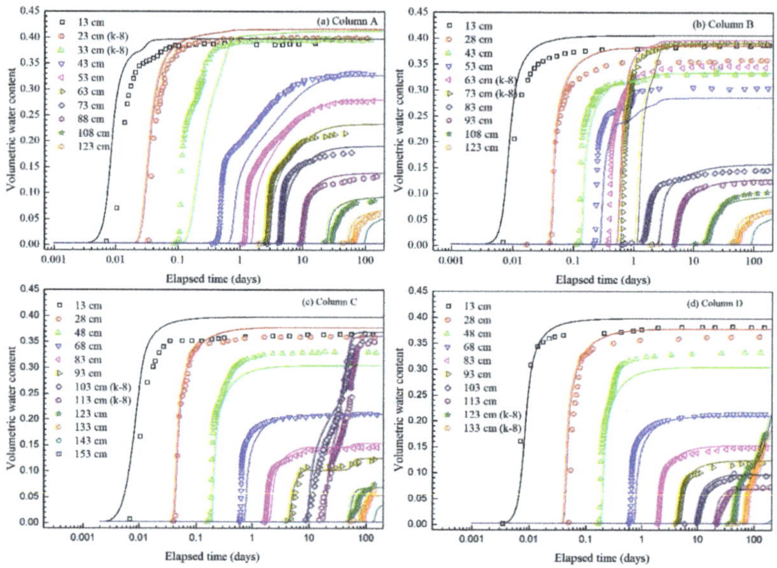

Figure 7. Comparison of the water content distribution in the layered soil profiles obtained by the test and the numerical simulation.

5. Conclusions

A series of capillary rise tests were conducted on soil columns of sand over fine sand over sand to investigate the effect of a relatively fine interlayer. In the layered soil, the water dynamics were affected not only by the interlayer properties and the thickness of the layers, but also by their spatial configuration. The soil water potential and the soil water distributions at equilibrium in the layered soil profile when the capillary water crossed the interlayer were different from those of the homogeneous soil column. The results showed that the water content was not continuous in the entire soil profile, and there were jumps at the coarser–finer soil and finer–coarser soil interfaces. The distribution of the matric suctions (calculated from the soil water retention curves) in this layered soil was similar to a profile without layering, but not the same. There was a small gap between these two wherever the finer layer in the middle was laid.

In the research work, the surface characteristics of soil particle such as the hydrophobicity was not taken into consideration. According to one of the reviewer's oping, the oven method can cause the possibility of hydrophobicity of the soil; however, the extent of the hydrophobicity should be studied in the further research. Although the column experiments had been conducted twice, only the successful case was presented in the paper, the other was failed due to the underestimation of the maximum capillary height of the homogeneous k-8 soil. In engineering practice, using the coarser layer over finer layer can cut off the capillary water, which may take the salt rise to higher position in soil layer and cause the corrosion of concrete.

Author Contributions: Conceptualization, Z.L. and Z.Z.; Investigation, H.L. (Haitao Li) and H.S.; Writing—original draft preparation, Q.L.; Writing—review and editing, Z.Z.; Project administration, H.L. (Haiyan Liu); funding acquisition, Q.L. All authors have read and agreed to the published version of the manuscript.

Funding: This research was supported in part by Shandong Provincial Natural Science Foundation, China (ZR2021MD021).

Institutional Review Board Statement: Not applicable.

Informed Consent Statement: Not applicable.

Conflicts of Interest: The authors declare no conflict of interest.

References

1. Xu, X.; Sun, C.; Qu, Z.; Huang, Q.; Ramos, T.B.; Huang, G. Groundwater recharge and capillary rise in irrigated areas of the upper yellow river basin assessed by an agro-hydrological model. *Irrig. Drain.* **2016**, *64*, 587–599. [CrossRef]
2. Hird, R.; Bolton, M.D. Clarification of capillary rise in dry sand. *Eng. Geol.* **2017**, *230*, 77–83. [CrossRef]
3. Polansky, J.; Kaya, T. An experimental and numerical study of capillary rise with evaporation. *Int. J. Therm. Sci.* **2015**, *91*, 25–33. [CrossRef]
4. Liu, Q.; Yasufuku, N.; Miao, J.; Ren, J. An approach for quick estimation of maximum height of capillary rise. *Soils Found.* **2014**, *54*, 1241–1245. [CrossRef]
5. Wösten, J.H.M.; Pachepsky, Y.; Rawls, W. Pedotransfer functions: Bridging the gap between available basic soil data and missing soil hydraulic characteristics. *J. Hydrol.* **2001**, *251*, 123–150. [CrossRef]
6. Bai, B.; Jiang, S.; Liu, L.; Li, X.; Wu, H. The transport of silica powders and lead ions under unsteady flow and variable injection concentrations. *Powder Technol.* **2021**, *387*, 22–30. [CrossRef]
7. Bai, B.; Nie, Q.; Wu, H.; Hou, J. The attachment-detachment mechanism of ionic/nanoscale/microscale substances on quartz sand in water. *Powder Technol.* **2021**, *394*, 1158–1168. [CrossRef]
8. Bai, B.; Wang, Y.; Rao, D.; Bai, F. The effective thermal conductivity of unsaturated porous media deduced by pore-scale sph simulation. *Front. Earth Sci.* **2022**, *10*, 943853. [CrossRef]
9. Zhu, J.; Mohanty, B.P. Analytical solutions for steady state vertical infiltration. *Water Resour. Res.* **2002**, *38*, 20-21–20-25. [CrossRef]
10. Huang, M.B.; Barbour, S.; Elshorbagy, A.; Zettl, J.; Si, B. Infiltration and drainage processes in multi-layered coarse soils. *Can. J. Soil Sci.* **2011**, *91*, 169–183. [CrossRef]
11. Ma, Y.; Feng, S.; Zhan, H.; Liu, X.; Su, D.; Kang, S.; Song, X. Water infiltration in layered soils with air entrapment: Modified green-ampt model and experimental validation. *J. Hydrol. Eng.* **2011**, *16*, 628–638. [CrossRef]
12. Serrano, S.E. Modeling infiltration in hysteretic soils. *Adv. Water Resour.* **1990**, *13*, 12–23. [CrossRef]
13. Bloemen, G. Calculation of steady state capillary rise from the groundwater table in multi-layered soil profiles. *Z. Pflanz. Bodenkd.* **1980**, *143*, 701–719. [CrossRef]
14. Shokri, N.; Lehmann, P.; Or, D. Evaporation from layered porous media. *J. Geophys. Res. Solid Earth* **2010**, *115*, 258–273. [CrossRef]
15. Japanese Geotechnical Society. *Laboratory Testing Standards of Geomaterials*; Japanese Geotechnical Society: Tokyo, Japan, 2017.
16. *BS EN ISO 14688-2*; Geotechnical Investigation and Testing Identification and Classification of Soil, Part 2: Principles for a Classification. British Standards Instituation: London, UK, 2004.
17. Amer, A.M.; Awad, A. Permeability of cohesionless soils. *J. Geotech. Eng. Div. ASCE* **1974**, *100*, 1309–1316. [CrossRef]
18. Van Genuchten, M.T. A closed-form equation for predicting the hydraulic conductivity of unsaturated soils. *Soil Sci. Soc. Am. J.* **1980**, *44*, 892–898. [CrossRef]
19. Van Genuchten, M.T.; Simunek, J.; Leij, F.; Sejna, M. *The Retc Code for Quantifying the Hydraulic Functions of Unsaturated Soils, Version 1.0.*; Epa Report 600/2-91/065; U.S. Salinity Laboratory, USDA, ARS: Riverside, CA, USA, 1991.
20. Mualem, Y. A new model for predicting the hydraulic conductivity of unsaturated porous media. *Water Resour. Res.* **1976**, *12*, 593–622. [CrossRef]
21. Šimůnek, J.; van Genuchten, M.; Šejna, M. Development and applications of the hydrus and stanmod software packages and related codes. *Vadose Zone J.* **2008**, *7*, 587–600. [CrossRef]
22. Richards, L.A. Capillary conduction of liquids through porous mediums. *Physics* **1931**, *1*, 318–333. [CrossRef]

applied sciences

MDPI

Article

Stability Analysis of Filled-Slope Reinforced by Frame with Prestressed Anchor-Plates under Static Action

Jun Zhang [1], Weili Li [2,*] and Shuaihua Ye [1,*]

1 School of Civil Engineering, Lanzhou University of Technology, Lanzhou 730050, China
2 GanSu CSCEC Municipal Engineering Investigation And Design Institute Co., Ltd., Lanzhou 730000, China
* Correspondence: csceclwl@163.com (W.L.); yeshuaihua@163.com (S.Y.)

Abstract: Because of the current situation where the stability research of filled-slope reinforced by a frame with prestressed anchor-plates lags behind the actual engineering application, based on the ultimate balance theory, the calculation formulas of stability factor under the four arc slip surface of filled-slopes reinforced by a frame with prestressed anchor-plates are derived by using the improved Bishop method; the corresponding search method of the most dangerous slip surface is given and the calculation formulas of the pullout force of anchor-plates are improved. Based on two examples, the stability results calculated by the proposed algorithm are compared with those calculated by PLAXIS 3D and GeoStudio 2012 finite element software, and the following conclusions are drawn. (1) The improved pullout force of anchor-plates takes into account the friction of the front and rear surface of the anchor-plate and the effect of cohesion of fill soil in the passive earth pressure on the front end of the anchor-plate, which makes the force of the anchor-plate more complete. (2) The stability factor of example 1 calculated by this method differs from the results simulated by PLAXIS 3D and GeoStudio 2012 by 4.6% and 7.1%, respectively; the stability factor of example 2 calculated by this method differs from the results simulated by PLAXIS3D and GeoStudio 2012 by 3.2% and 4.5%, respectively, which can meet the engineering requirements. (3) The stability analysis method of filled-slope reinforced by a frame with prestressed anchor-plates that is proposed is reasonable and suitable for any arc slip surface in the filled-slope reinforced by a frame with prestressed anchor-plates, and it provides some guiding values for the design of practical engineering.

Keywords: filled-slope; static action; frame with prestressed anchor-plates; stability

check for updates

Citation: Zhang, J.; Li, W.; Ye, S. Stability Analysis of Filled-Slope Reinforced by Frame with Prestressed Anchor-Plates under Static Action. *Appl. Sci.* **2023**, *13*, 1615. https://doi.org/10.3390/app13031615

Academic Editor: Bing Bai

Received: 11 December 2022
Revised: 6 January 2023
Accepted: 18 January 2023
Published: 27 January 2023

1. Introduction

In recent years, a large number of cut slopes and filled-slopes have often been produced in the process of urbanization construction. In order to prevent the occurrence of landslides, appropriate supporting structures must be adopted to reinforce the slopes [1–6]. In the project, gravity retaining walls [7,8], cantilever retaining walls [9,10], counterfort retaining walls [11,12] and anchor slab retaining walls [13,14] are generally used to reinforce filled-slopes. However, when the slope exceeds 15 m, if the above methods are used, the reinforcement effect is not ideal, the construction is difficult and the economy is unreasonable, and then there are some hidden dangers. Technical code for building slope engineering (GB 50330-2013) [15] stipulates that special designs should be carried out for soil slopes above 15 m, and effective and reliable strengthening measures should be taken. To this end, Ye and Zhu [16,17] proposed the frame with prestressed anchor-plates suitable for filled-slopes. This structure not only overcomes the height limitation of the above filled-slope supporting structure, but also has the advantages of convenient construction, low cost, and good overall stability. Plus, the deformation of the filled-slope can be well controlled by applying prestress to the anchor-plates. At present, this structure has been applied to the practical project; Figure 1 is the scene picture of the anchor-plate, and Figure 2 is a filled-slope reinforced by a frame with prestressed anchor-plates.

Figure 1. The scene picture of the anchor-plate.

Figure 2. Filled-slope reinforced by a frame with prestressed anchor-plates.

As a new type of flexible supporting structure, the structural form and working mechanism of frames with prestressed anchor-plates are similar to those of frame prestressed anchor supporting structures and anchor slab retaining walls. Under the action of the supporting structure, the soil properties of the original slope will change, and the deformation and interface characteristics of the soil material are different from the original slope, which will have a great impact on the stability of the slope [18–21]. Therefore, the slope stability analysis considering the action of the supporting structure is also the core content in the field of slope research. The current research on the stability of slopes reinforced by a frame prestressed anchor and anchor slab retaining wall under static action are as follows. Li et al. [22] established an intelligent optimization calculation model of anchor tension at each layer that satisfies the excavation process stability and realized the real-time dynamic analysis of stability of slopes reinforced by a frame prestressed anchor in the process of excavation and reinforcement. Zhu et al. [23], based on the basic principle of the upper limit theory of plastic mechanics, the safety coefficient calculation formula of a slope reinforced by a frame with prestressed anchors is derived. He et al. [24] proposed a stability analysis method of a suspended-anchor earth retaining wall. The current research on filled-slopes reinforced by a frame with a prestressed anchor-plate supporting structure are as follows. Ye et al. [16] compared numerical simulation results with the actual monitoring data of a filled-slope reinforced by the frame with prestressed anchor-plates, and found that this supporting structure can effectively improve the stability of the filled-slope and can effectively control the displacement and deformation of the filled-slope. Zhu et al. [25], based on the ultimate balance theory, the stability calculation method of a filled-slope reinforced by a frame with prestressed anchor-plates is proposed by using the circular vertical simple slice method, and the ultimate bearing capacity calculation formula of anchor-plates is given. In addition to the above theoretical research on slope stability with a supporting structure, when using finite element software to analyze the stability of a filled-slope reinforced by a frame with prestressed anchor-plates, the simulation of anchor-plates is also worthy of attention. Phung et al. [26,27] used finite element software to simulate different types of plates and analyzed the simulation results, which provided a certain reference for the simulation of anchor-plates.

From the present situation of the above research, it can be found that the theoretical research and numerical simulation research on the stability of filled-slopes reinforced by a frame with prestressed anchor-plates is very immature. Thus, this paper proposes four types of slip surfaces that may occur when the frame with prestressed anchor-plates is used to reinforce the filled-slope. It is divided into two types according to whether there is the lowest point F of the slip surface. Based on the ultimate balance theory, the improved Bishop method is used to deduce the stability calculation formula under two different types of slip surfaces, the corresponding search model of the most dangerous slip surface is given, and the calculation theory of pullout force of anchor-plates is improved. The two kinds of stability algorithms are optimized, and the stability analysis flow of a filled-slope reinforced by a frame with a prestressed anchor-plate supporting structure is given, which makes it more applicable. Based two examples, the stability results calculated by the proposed algorithm are compared with those calculated by PLAXIS 3D finite element software and GeoStudio 2012 finite element software to verify the rationality of the proposed stability analysis method of a filled-slope reinforced by a frame with prestressed anchor-plates. This method is suitable for any arc slip surface that appears on the filled-slope reinforced by the frame with prestressed anchor-plates, and it provides some guiding value for the design of practical engineering.

2. Composition and Working Mechanism of Frame with Prestressed Anchor-Plate Supporting Structure

The frame with a prestressed anchor-plate supporting structure is composed of a reinforced concrete frame and an anchor-plate, which belongs to a kind of soft filled-slope supporting structure in the geotechnical anchoring structure system. Its elevation and section drawings are shown in Figures 3 and 4. The anchor-plate consists of a tie rod and a reinforced concrete prefabricated panel. When the filled-slope is reinforced, the structural system is like a floor well-shaped beam structure standing on the soil, the prestressed anchor-plate is anchored in the soil of the slope, and the anchor head of the anchor-plate is connected to the node of the frame beam. The earth pressure on the frame beam is finally transferred to the anchor-plate located in the stable area through the pull rod, and then the slope stability is maintained by the friction between the anchor-plate and the surrounding soil.

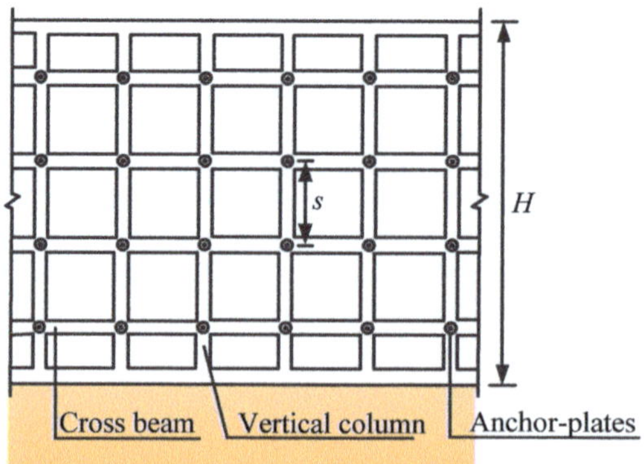

Figure 3. Elevation view of a frame with a prestressed anchor-plate supporting structure.

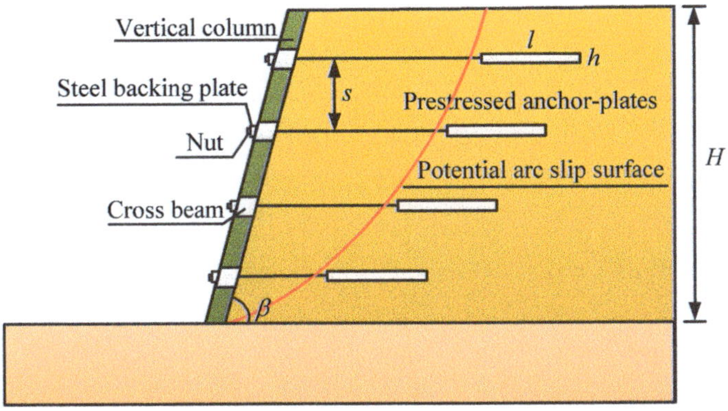

Figure 4. Section view of a frame with a prestressed anchor-plate supporting structure.

3. Types of Slip Surfaces That May Appear in The Reinforcement of The Filled-Slope by the Frame with Prestressed Anchor-Plates

Based on the failure of soil slopes, the filled-slope reinforced by a frame with prestressed anchor-plates may actually appear in the following four types of slip surfaces.

(1) When the arc slip surface passes over the toe of the slope, it is called the slope toe circle [28,29]. There are two main forms of slope toe circle, one of which is shown in Figure 5. The angle between the tangent of any point on the slip plane and the horizontal direction is between 0°~90°, and the center P of the slip plane is located at the upper left of the whole slope. There is much research on slope stability in the case of this type of slip surface.

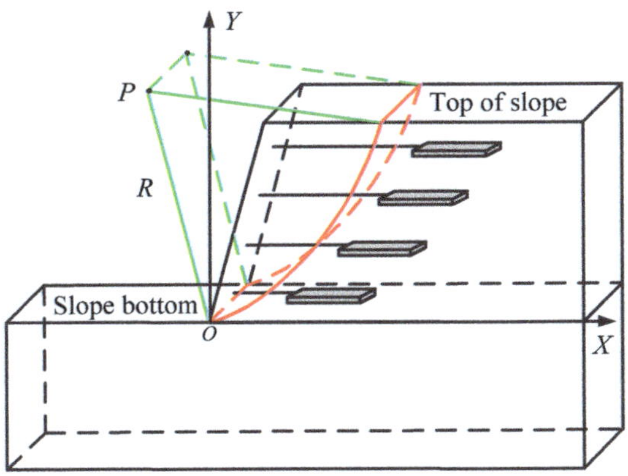

Figure 5. Schematic diagram of slope toe circle 1.

(2) Another form of the slope toe circle is shown in Figure 6. Although this slip surface passes over the toe of the slope, the angle between the tangent and the horizontal direction at the toe of the slope is a negative angle, and the lowest point F on the arc slip surface is on the right side of the toe of the slope.

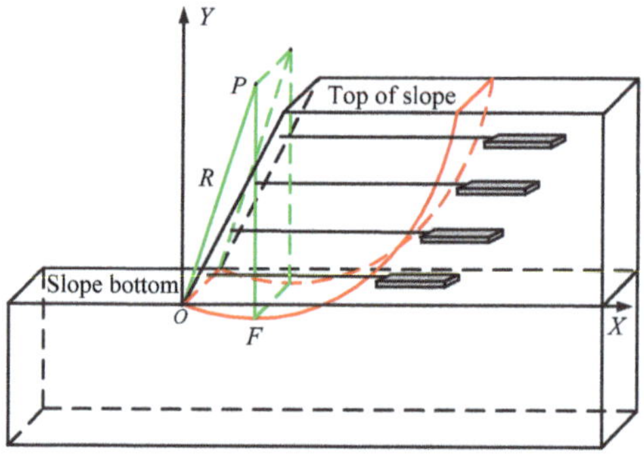

Figure 6. Schematic diagram of slope toe circle 2.

(3) When the arc slip surface passes through a certain position other than the toe of the slope, it is called the midpoint circle [28,29]. There are also two main forms of midpoint circles, one of which is shown in Figure 7. The lowest point F and the center P of the arc slip surface are both located on the right side of the toe of the slope.

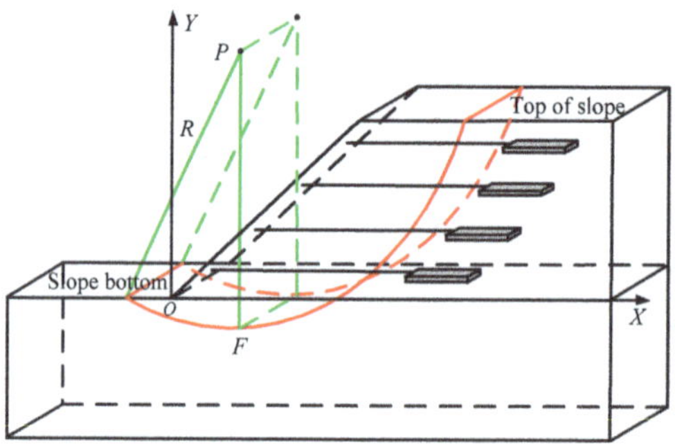

Figure 7. Schematic diagram of slope toe circle 2.

(4) Another form of the midpoint circle is shown in Figure 8. The lowest point F and center P of the arc slip surface are located on the left side of the toe of the slope, and the sliding range of this slip surface is larger than that shown in Figure 7.

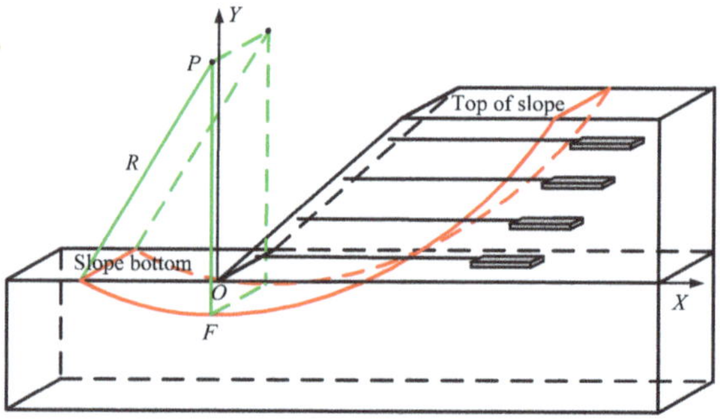

Figure 8. Schematic diagram of midpoint circle 1.

Among the four types of slip surfaces mentioned above, they are divided into two types according to whether there is the lowest point F. The slope toe circle 1 is classified as the first type slip surface, and the slope toe circle 2, the midpoint circle 1 and the midpoint circle 2 are classified as the second type slip surface.

4. Stability Analysis Model of Filled-Slope Reinforced by Frame with Prestressed Anchor-Plates

The simplified Bishop method, as one of the ultimate balance methods, is considered to be a "non-strict" method because it ignores the shear force between strips and does not strictly satisfy the equilibrium conditions. However, many scholars have found that the stability factors of slopes obtained by the improved Bishop method and the strict method are very close, which can be called the strict strip method [30–32]. The traditional simplified Bishop method is only suitable for the stability analysis of natural slope, but cannot be directly used to analyze the stability of a slope with a supporting structure. Therefore, considering the influence of a frame with a prestressed anchor-plate supporting structure on the stability of a filled-slope, this paper improves the traditional simplified Bishop method and deduces the stability calculation formula which is suitable for a filled-slope reinforced by a frame with prestressed anchor-plates.

4.1. Basic Assumptions

According to the central idea of the Bishop method, the following assumptions are made [30]:

(1) The slope is a soil slope, and the slip surface is an arc slip surface.
(2) The shear strength of slope soil obeys Mohr–Coulomb criterion.
(3) The inter-strip shear force is ignored, and the inter-strip horizontal force is considered.
(4) The effect of the anchor-plate on the slope is equivalent to the force perpendicular to the tangent direction of the slip plane, and this force is uniformly distributed on the slip surface.

4.2. Stability Analysis in the Case of the First Type of Slip Surface

4.2.1. Solving the Stability Factor

The diagram of the stability analysis for the first type of slip surface is shown in Figure 9.

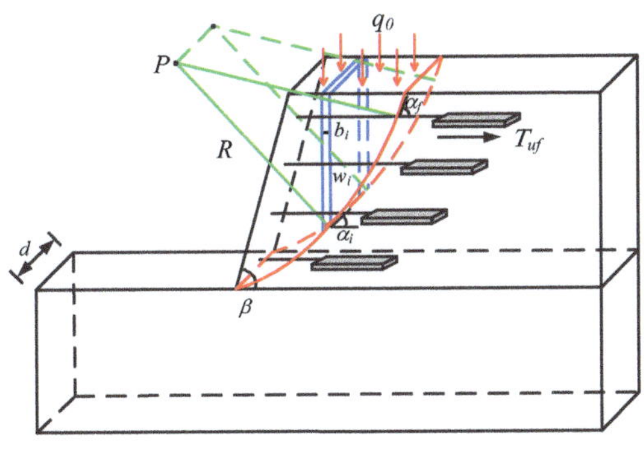

Figure 9. Diagram of stability analysis for the first type of slip surface.

When analyzing the stability of the slope shown in Figure 9 based on the simplified Bishop method, due to the supporting effect of the frame with prestressed anchor-plates on the slope, the force of the soil strip is different from that of the soil strip without reinforcement. Thus, the resistant shear force provided by the anchor-plates and the resistant shear force of the soil itself cannot be simply superimposed. As shown in Figure 10, T_u' and T_u'' can be obtained by decomposing the ultimate bearing capacity provided by anchor-plates. T_u' and soil have the same direction of resistant shear force of the soil, which directly provides resistant shear force for the sliding body. T_u'' is perpendicular to T_u', and it provides resistant shear force for the sliding body through friction with the soil outside the sliding surface. In order to derive the formula for calculating the slope stability under the action of the supporting structure, T_u' is equivalent to the force in the same direction as T_u'' according to Equation (2). The force provided by the anchor-plates is uniformly distributed on the slip surface as the force perpendicular to the tangent direction of the slip surface, so the force provided by the anchor-plates on the soil strip i is as follows:

$$N_{ui} = [\sum_{f=1}^{t} (T_{uf} \sin \alpha_f + T_{uf} \cos \alpha_f F_s / \tan \varphi_f)] b_i \sec \alpha_i / L \tag{1}$$

where t is the total number of layers of the prestressed anchor-plate; T_{uf} is the ultimate bearing capacity of the f-layer anchor-plate, which can be determined by Equation (31); α_f is the angle between the tangent line and the horizontal direction at the intersection of the f-layer anchor-plate and the slip surface; φ_f is the internal friction angle of the soil at the intersection point of the f-layer anchor-plate and the slip surface; L is the total arc length of the slip surface.

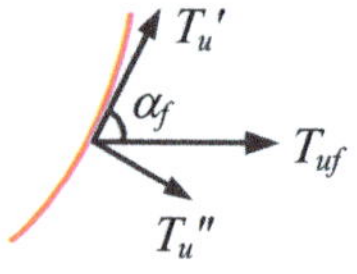

Figure 10. Resolution diagram of the ultimate bearing capacity of the anchor-plate.

The force diagram of the soil strip considering the action of the supporting structure is shown in Figure 11, according to the Mohr–Coulomb criterion [30]:

$$T_i d = N_i d \tan \varphi_i / F_s + c_i b_i d \sec \alpha_i / F_s \tag{2}$$

where F_s is the stability factor of slope; $T_i d$ is the shear force at the bottom of the soil strip i; $N_i d$ is the normal force at the bottom of the soil strip i; α_i is the angle between the tangent line and the horizontal direction at the slip surface of the soil strip i; c_i is the cohesion at the slip surface of the soil strip i; φ_i is the angle of internal friction at the slip surface of the soil strip i; b_i is the width of the soil strip i; d is the thickness of sliding element.

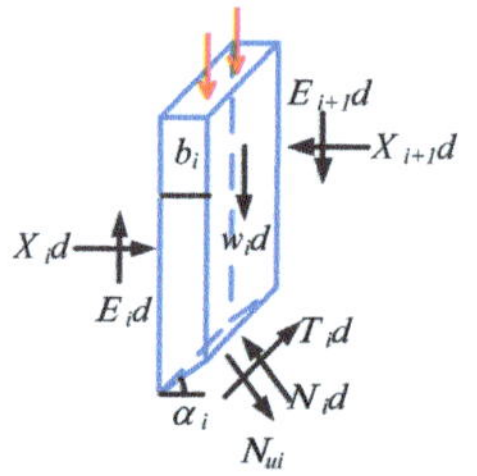

Figure 11. Force diagram of soil strip i.

Considering the vertical force balance [30]:

$$(N_i d - N_{ui}) \cos \alpha_i + T_i d \sin \alpha_i = (W_i + q_0 b_i) d \tag{3}$$

where $W_i d$ is the dead weight of the soil strip i; q_0 is the uniformly distributed load on the top of the slope.

Simultaneous Equations (2) and (3) can be obtained:

$$N_i d = [(W_i + q_0 b_i) d + N_{ui} \cos \alpha_i - c_i b_i d \tan \alpha_i / F_s] / m_{\alpha i} \tag{4}$$

$$T_i d = [(W_i + q_0 b_i) d \tan \varphi_i / F_s + N_{ui} \cos \alpha_i \tan \varphi_i / F_s + c_i b_i d / F_s] / m_{\alpha i} \tag{5}$$

$$m_{\alpha i} = \cos \alpha_i + \sin \alpha_i \tan \varphi_i / F_s \tag{6}$$

Consider the moment balance to the center of the circle:

$$\sum_{i=1}^{n} T_i dR = \sum_{i=1}^{n} (W_i + q_0 b_i) d \sin \alpha_i R \tag{7}$$

where n is the number of the strips of sliding body; R is the arc radius of the slip surface.
Substitute Equation (5) into Equation (7) to obtain:

$$F_s = \frac{\sum_{i=1}^{n} [(W_i + q_0 b_i) d \tan \varphi_i + c_i b_i d + N_{ui} \cos \alpha_i \tan \varphi_i] R / m_{\alpha i}}{\sum_{i=1}^{n} (W_i + q_0 b_i) d \sin \alpha_i R} \tag{8}$$

Substitute Equation (1) into Equation (8) to obtain:

$$F_s = \frac{\sum_{i=1}^{n} [(W_i + q_0 b_i) d \tan \varphi_i + c_i b_i d + \sum_{f=1}^{t} (T_{uf} \sin \alpha_f + T_{uf} \cos \alpha_f F_s / \tan \varphi_f) b_i \tan \varphi_i / L] R / m_{\alpha i}}{\sum_{i=1}^{n} (W_i + q_0 b_i) d \sin \alpha_i R} \tag{9}$$

4.2.2. Search for Models of Slip Surfaces

(1) The center coordinates and radius of the slip surface

The search model of the first kind of slip surface is shown in Figure 12. Although the established model is a three-dimensional model, the slip surface always changes in the *xoy* plane, so it is simplified to two-dimensional coordinates. In the Cartesian coordinate system, $P(x_c, y_c)$, $C(e, H)$, $O(0, 0)$, the angle between the tangent of point O and the horizontal is θ, the slope of the tangent is k', $k' = \tan\theta$.

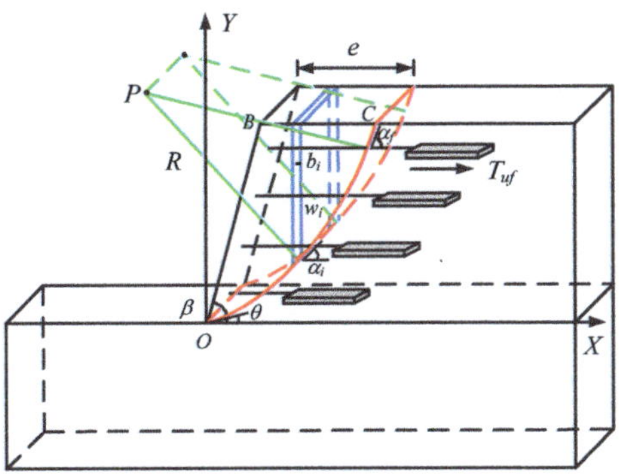

Figure 12. Search model of the first kind of slip surface.

Then the slip surface equation is:

$$(x - x_c)^2 + (y - y_c)^2 = R^2 \tag{10}$$

And satisfies:

$$\left. \begin{array}{l} x_c{}^2 + y_c{}^2 = R^2 \\ (e - x_c)^2 + (H - y_c)^2 = R^2 \\ x_c = -k' y_c \end{array} \right\} \tag{11}$$

Combining Equations (10) and (11) yields:

$$\left.\begin{array}{l} x_c = \frac{-k'(H^2+e^2)}{2H-2k'e} \\ y_c = \frac{H^2+e^2}{2H-2k'e} \\ R = \frac{\sqrt{1+k'^2}(H^2+e^2)}{2H-2k'e} \end{array}\right\} \tag{12}$$

Therefore, the slip surface is controlled by θ and e. The value range of θ and e is determined, and the appropriate step size is set respectively, and the search of slip surface can be realized by constantly changing.

(2) The angle between the tangent of any point on the arc and the horizontal plane

At any point $M(x_k, y_k)$ on the arc, the angle between the tangent and the horizontal plane is α_k, which can be obtained from the geometric relationship:

$$\sin \alpha_k = \frac{x_c - x_k}{R} \tag{13}$$

$$\cos \alpha_k = \frac{y_c - y_k}{R} \tag{14}$$

(3) The self-weight of the soil block i

According to Figure 9, the coordinate of the soil strip i at the slip surface is (x_i, y_i), and when the upper end of the soil strip is on OB, the upper coordinate is (x_{i1}, y_{i1}), so the formula for calculating the self-weight of the soil strip i is as follows:

$$W_i = \begin{cases} \gamma b_i(H - y_i)\ x_i \geq \frac{H}{\tan\beta} \\ \gamma b_i(y_{i1} - y_i)\ 0 \leq x_i \leq \frac{H}{\tan\beta} \end{cases} \tag{15}$$

where γ is the weight of fill soil.

4.3. *Stability Analysis in the Case of the Second Type of Slip Surface*

4.3.1. Solving the Stability Factor

The diagram of the stability analysis for the second type of slip surface and the force diagram of soil strip j are shown in Figures 13 and 14.

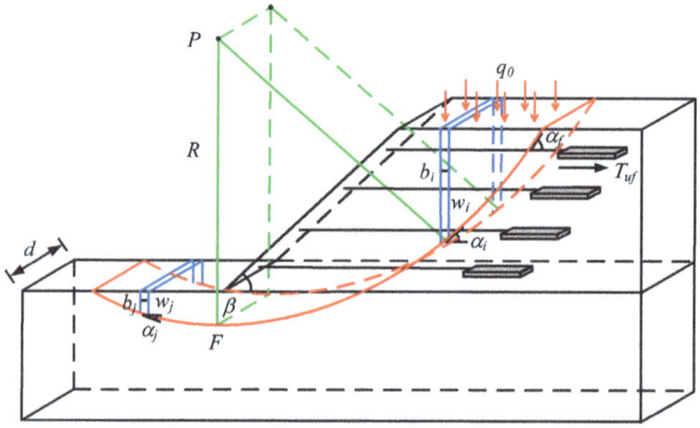

Figure 13. Diagram of stability analysis for the second type of slip surface.

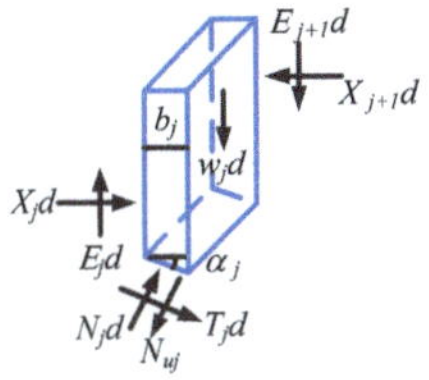

Figure 14. Force diagram of soil strip *j*.

Soil strip on the left side of *OF*.

The force provided by the anchor-plates on the soil strip *j* is as follows:

$$N_{uj} = [\sum_{f=1}^{t} (T_{uf} \sin \alpha_f + T_{uf} \cos \alpha_f F_s / \tan \varphi_f)] b_j \sec \alpha_j / L \quad (16)$$

By Mohr–Coulomb criterion [30]:

$$T_j d = N_j d \tan \varphi_j / F_s + c_j b_j d \sec \alpha_j / F_s \quad (17)$$

where T_jd is the shear force at the bottom of the soil strip *j* at the left on point *F*; N_jd is the normal force at the bottom of the soil strip *j* at the left on point *F*; b_j is the width of the soil strip *j* at the left on point *F*; c_j is the cohesion at the slip surface of the soil strip *j* at the left on point *F*; φ_j is the angle of internal friction at the slip surface of the soil strip *j* at the left on point *F*; α_j is the angle between the tangent line and the horizontal plane at the slip plane of the soil strip *j* at the left on point *F*.

Consider the vertical force balance [30]:

$$(N_j d - N_{uj}) \cos \alpha_j - T_j d \sin \alpha_j = W_j d \quad (18)$$

where W_jd is the dead weight of the soil strip *j* at the left on point *F*.

Simultaneous Equations (17) and (18) can be obtained:

$$N_j d = (W_j d + N_{uj} \cos \alpha_j + c_j b_j \tan \alpha_j / F_s) / m_{\alpha j} \quad (19)$$

$$T_j d = (W_j d \tan \varphi_j / F_s + N_{uj} \cos \alpha_j \tan \varphi_j / F_s + c_j b_j d / F_s) / m_{\alpha j} \quad (20)$$

$$m_{\alpha j} = \cos \alpha_j - \sin \alpha_j \tan \varphi_j / F_s \quad (21)$$

Soil strip on the right side of *OF*.

The force analysis of the soil strip on the right side of *OF* is similar to that in the case of slip surface 1, and the final deduced results are Equations (4)–(6).

Consider the moment balance to the center of the circle:

$$\sum_{i=1}^{n} T_i d R + \sum_{j=1}^{m} T_j d R = \sum_{i=1}^{n} (W_i + q_0 b_i) d \sin \alpha_i R - \sum_{j=1}^{m} W_j d \sin \alpha_j R \quad (22)$$

where *n* is the number of the strips of sliding body at the right on point *F*; *m* is the number of the strips of sliding body at the left on point *F*; *R* is the arc radius of the slip surface; T_id is the shear force at the bottom of the soil strip *i* at the right on point *F*; W_id is the dead weight of the soil strip *i* at the right on point *F*; α_i is the angle between the tangent line and the horizontal plane at the slip surface of the soil strip *i* at the right on point *F*.

Substitute Equations (5) and (20) into Equation (22) to obtain:

$$F_s = \frac{\sum\limits_{i=1}^{n} [(W_i + q_0 b_i) d \tan\varphi_i + c_i b_i d + N_{ui}\cos\alpha_i \tan\varphi_i] R / m_{\alpha i} + \sum\limits_{j=1}^{m} [(W_j d \tan\varphi_j + c_j b_j d + N_{uj}\cos\alpha_j \tan\varphi_j] R / m_{\alpha j}}{\sum\limits_{i=1}^{n} (W_i + q_0 b_i) d \sin\alpha_i R - \sum\limits_{j=1}^{m} W_j d \sin\alpha_j R} \tag{23}$$

where b_i = the width of the soil strip i at the right on point F; c_i = the cohesion at the slip surface of the soil strip i at the right on point F; φ_i = the angle of internal friction at the slip surface of the soil strip i at the right on point F.

Substitute Equations (1) and (16) into Equation (23) to obtain:

$$F_s = \frac{\sum\limits_{i=1}^{n} [(W_i + q_0 b_i) d \tan\varphi_i + c_i b_i d + \sum\limits_{f=1}^{t} (T_{uf}\sin\alpha_f + T_{uf}\cos\alpha_f F_s / \tan\varphi_f) b_i \tan\varphi_i / L] R / m_{\alpha i} + \sum\limits_{j=1}^{m} [W_j d \tan\varphi_j + c_j b_j d + \sum\limits_{f=1}^{t} (T_{uf}\sin\alpha_f + T_{uf}\cos\alpha_f F_s / \tan\varphi_f) b_j \tan\varphi_j / L]) R / m_{\alpha j}}{\sum\limits_{i=1}^{n} (W_i + q_0 b_i) d \sin\alpha_i R - \sum\limits_{j=1}^{m} W_j d \sin\alpha_j R} \tag{24}$$

4.3.2. Search for Models of Slip Surfaces

(1) The center coordinates and radius of the slip surface

The search model of the second kind of slip surface is shown in Figure 15. Since the slip surface always changes in the *xoy* plane, it is simplified to two-dimensional coordinates. In the Cartesian coordinate system, $P(x_c, y_c)$, $C(e, H)$, $A(-e_0, 0)$, $O(0, 0)$, the angle between the tangent of point A and the horizontal is θ, and the slope of the tangent is k', $k' = -tan\theta$.

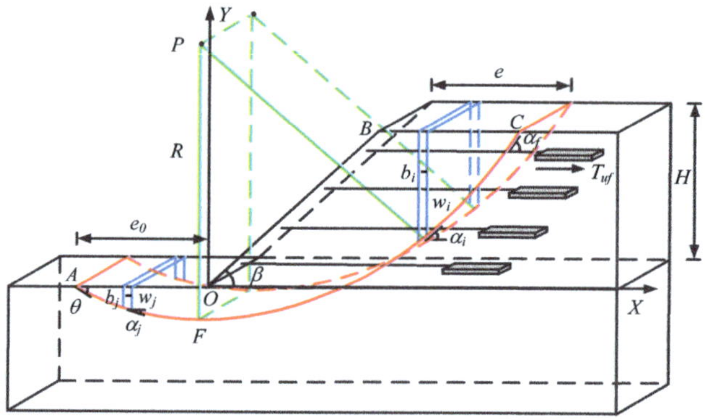

Figure 15. Search model of the second kind of slip surface.

Then the slip surface equation is:

$$(x - x_c)^2 + (y - y_c)^2 = R^2 \tag{25}$$

And satisfies:

$$\left.\begin{array}{l}(-e_0 - x_c)^2 + {y_c}^2 = R^2 \\ (e - x_c)^2 + (H - y_c)^2 = R^2 \\ x_c = -(k' y_c + e_0)\end{array}\right\} \tag{26}$$

Combining Equations (25) and (26) yields:

$$\left.\begin{array}{l} x_c = -e_0 - \frac{k'[(e+e_0)^2+H^2]}{2H-2k'(e+e_0)} \\ y_c = \frac{(e+e_0)^2+H^2}{2H-2k'(e+e_0)} \\ R = \frac{\sqrt{1+k'^2}[(e+e_0)^2+H^2]}{2H-2k'(e+e_0)} \end{array}\right\} \tag{27}$$

Therefore, the slip surface is controlled by θ, e_0 and e. The value range of θ, e_0 and e is determined, and the appropriate step size is set respectively, and the search of slip surface can be realized by constantly changing.

(2) The angle between the tangent of any point on the arc and the horizontal plane

At any point $M'(x_k, y_k)$ on the arc, the angle between the tangent and the horizontal plane is α_k, which can be obtained from the geometric relationship:

$$\sin\alpha_k = \frac{|x_c - x_k|}{R} \tag{28}$$

$$\cos\alpha_k = \frac{y_c - y_k}{R} \tag{29}$$

(3) The self-weight of the soil block i and j

It can be seen from Figure 12 that the calculation of the self-weight of the soil strip under the second type of slip surface is more complicated. The coordinates of the soil strip i at the slip surface are (x_i, y_i). When the upper end of the soil strip i is on the OB, its upper coordinate is (x_{i1}, y_{i1}). The coordinates of the soil strip j at the slip plane are (x_j, y_j). When the upper end of the soil strip j is on OB, its upper coordinate is (x_{j1}, y_{j1}). When the abscissa of the zero-boundary point of the slip surface crossing from the filling layer to the foundation soil layer is x_0, the calculation formulas of the self-weight of the soil strip i and j are as follows:

$$W_{i/j} = \begin{cases} \gamma_1 b_i(H - y_i) & x_i \geq \frac{H}{\tan\beta} \\ \gamma_1 b_i(y_{i1} - y_i) & x_0 \leq x_i < \frac{H}{\tan\beta} \\ \gamma_1 b_i y_{i1} - \gamma_2 b_i y_i & 0 \leq x_i < x_0 \\ \\ -\gamma_2 b_i y_i & x_c \leq x_i < 0 \\ -\gamma_2 b_j y_j & -e_0 \leq x_j < x_c \end{cases} \tag{30}$$

where γ_1 is the weight of fill soil; γ_2 is the weight of foundation soil.

4.4. The Ultimate Bearing Capacity of the Anchor-Plate

For the calculation of the ultimate bearing capacity of the anchor-plate, on the basis of the method in Zhu [25], the calculation method of the pullout resistance of the anchor-plate is improved. The force diagram of the anchor-plate is shown in Figure 16.

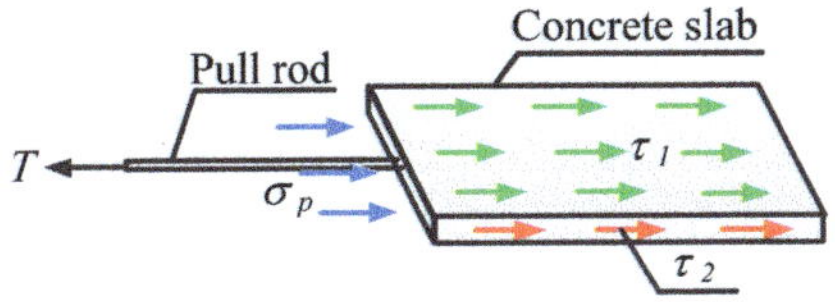

Figure 16. Force diagram of the anchor-plate.

According to Zhu [25], the formula for calculating the pullout force of the anchor-plate is as follows:

$$T_{uf} = \sum \min\{T_s, T_p\} \tag{31}$$

where T_{uf} is the ultimate bearing capacity of the f-layer anchor-plate; T_s is the ultimate tensile force of the steel rod; T_p is the limit value of the pullout force that the anchor-plate can provide, which is composed of the friction resistance of the interface between the anchor-plate and soil and the passive earth pressure of the anchor-plate.

$$T_s = f_y A_s \tag{32}$$

where f_y is the yield strength of the steel rod; A_s is the cross-sectional area of the steel rod.

$$T_p = F_f + Q \tag{33}$$

where F_f is the friction resistance on the anchor-plate; Q is passive earth pressure on the anchor-plate.

Because Zhu [25] did not consider the friction between the two sides of the anchor-plate and the soil when calculating the friction between the anchor-plate and the soil, this paper improves it, and the improved formula for calculating the friction between the anchor-plate and the soil is as follows:

$$F_f = 2(\tau_1 A_1 + \tau_2 A_2) \tag{34}$$

where τ_1 is the shear strength of the interface between the upper and lower interface of the anchor-plate and the soil; τ_2 is the shear strength of the interface between the front and rear interface of the anchor-plate and the soil; A_1 is the upper surface area of anchor-plate; A_2 is the side surface area of anchor-plate:

$$\left.\begin{array}{l} \tau_1 = \mu(\gamma h_0 + q_0) \\ \tau_2 = k_0 \mu(\gamma h_0 + q_0) \end{array}\right\} \tag{35}$$

where γ is the weight of the soil layer where the anchor-plate is located; h_0 is the height from the top of the slope to the soil layer where the anchor-plate is located; q_0 is the uniformly distributed load on the top of the slope; μ is the coefficient of lateral earth pressure:

$$Q = \sigma_p A_3 \tag{36}$$

where σ_p is the passive earth pressure of the soil layer where the anchor-plate is located; A_3 is the front-end area of anchor-plate:

$$\left.\begin{array}{l} A_1 = lb \\ A_2 = lh \\ A_3 = bh \end{array}\right\} \tag{37}$$

where l is the length of the anchor-plate; b = the width of the anchor-plate; h is the height of the anchor-plate.

Because Zhu [25] did not consider the cohesion of the fill soil when calculating the passive earth pressure on the front end of the anchor-plate, which obviously does not conform to the actual engineering situation, the improved formula is as follows:

$$\sigma_p = k_p(\gamma h_0 + q_0) + 2c\sqrt{k_p} \tag{38}$$

where c is the soil cohesion of the soil layer where the anchor-plate is located; k_p is the Rankine passive earth pressure coefficient of the soil around the anchor-plate.

4.5. Stability Analysis Process

Although this paper proposes four forms of slip surface for the reinforcement of a filled-slope with frame prestressed anchor-plates, it is also clear that the form of slip surface is related to the height of the slope, the degree of the slope, and soil conditions of the fill soil and the foundation soil. However, it is difficult to give clear certain conditions for the occurrence of various slip surfaces; as a result, it is not known which type of slip surface stability analysis process to select when an example is given. Therefore, the stability analysis process under the two types of slip surfaces mentioned above is optimized. The specific analysis process of the stability of the filled-slope reinforced by the frame prestressed anchor-plates is shown in Figure 17.

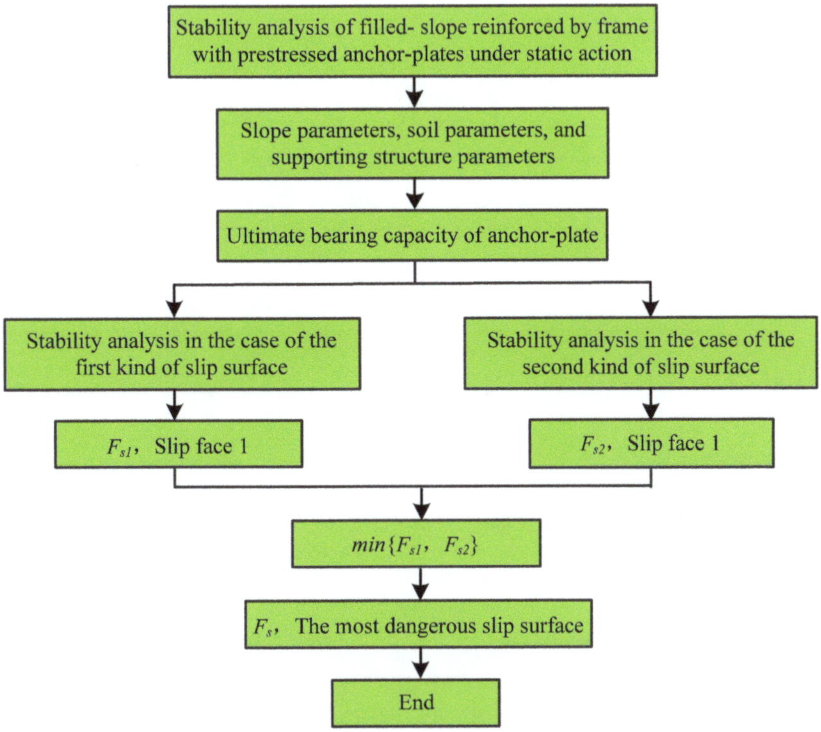

Figure 17. Flow chart of stability analysis.

5. Example Analysis

5.1. Example 1

The specific parameters of the filled-slope reinforced by the frame with prestressed anchor-plates in Lanzhou are as follows. The height of the slope is 12 m, the slope rate is 1: 0.5, the top load of the slope is q_0 = 20 kN/m^2, and the soil parameters are shown in Table 1. The horizontal spacing of the anchor-plate is 3 m, and the length, width and height of the anchor-plate are 4 m, 1 m, and 0.1 m, respectively, and the parameters of anchor-plates for each layer are shown in Table 2.

Table 1. Soil parameters of example 1.

Soil Layer Name	Natural Heavy γ/(kN/m^3)	Cohesive Force c/(kPa)	Angle of Internal Friction φ/(°)	Elastic Modulus E/(kN/m^2)	Friction Coefficient μ
Fill soil	17	20	24	35,000	0.4
Foundation soil	22	28	34	50,000	

Table 2. Parameters of anchor-plate of example 1.

Layer Number	Relative Ground Height/m	Length of Steel Tie Rod/m	Prestress Value/kN
Fourth floor	10.5	12.5	60
Third floor	7.5	11.0	90
Second floor	4.5	9.5	120
First floor	1.5	8.0	150

5.1.1. Analysis of Pullout Resistance of Anchor-Plate

The pullout force calculation method of anchor-plates in Zhu [25] and the improved method in this paper are used to calculate the pullout force of anchor-plates in example 1, and the results are shown in Table 3.

Table 3. Calculation results of pullout force of anchor-plate in example 1.

Layer Number	Original Method/kN	The Method of This Paper/kN	Percentage of Difference
Fourth floor	151.64	168.81	11.32%
Third floor	326.94	351.12	7.40%
Second floor	502.23	533.43	6.21%
First floor	677.52	715.74	6.08%

It can be seen from Table 3 that the pullout force of the anchor-plate calculated by this method is larger than that calculated by the method in Zhu [25]. This is because this method improves the method in Zhu [25]. From Equation (31) it can be seen that the pullout force of the anchor-plate is composed of the friction between the anchor-plate and the soil and the passive earth pressure on the front end of the anchor-plate, so the improvement of the calculation method of the pullout force of the anchor-plate in Zhu [25] is mainly reflected in two aspects. (1) When calculating the friction resistance between the anchor-plate and the surrounding soil, the friction resistance on the front and rear surface of the anchor-plate is considered in addition to the friction resistance on the upper and lower surface of the anchor-plate. Although the friction force on the front and rear surface of the anchor-plate is much smaller than that on the upper and lower surface, taking it into account will make the force of the anchor-plate more complete. (2) When calculating the passive earth pressure on the front end of the anchor-plate, the effect of soil cohesion is taken into account, which makes the calculation of the pullout force of the anchor-plate more suitable for practical engineering.

5.1.2. Stability Analysis

The stability calculation method of this paper, strength reduction method and ultimate balance method are used to calculate the stability of example 1, respectively.

(1) The method of this paper

The stability of example 1 is calculated by using the stability analysis process in this paper, and the calculated results are shown in Figure 18. It can be clearly seen from Figure 18 that the most dangerous slip surface passes over the toe of the slope, which belongs to the sliding mode of toe circle 1. The position of the most dangerous slip surface is 7 m away from point *B* at the top of the slope, and the calculated stability factor is 1.710.

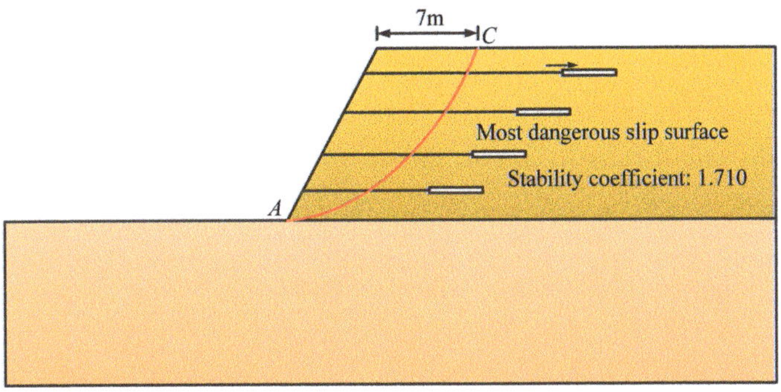

Figure 18. Example 1 calculation results of the stability of this method.

(2) Strength reduction method

PLAXIS 3D finite element software can establish a three-dimensional slope model and calculate its stability, so PLAXIS 3D finite element software is used to calculate the stability of example 1. The anchor-plate is simulated by a large area rectangular beam in the "Embedded beam" element, the frame beam is simulated by the "Beam" element, and the tension bar is simulated by the "Node to node anchor" element. The established three-dimensional model of the filled-slope with a slope ratio of 1:0.5 reinforced by the frame with prestressed anchor-plates is shown in Figure 19.

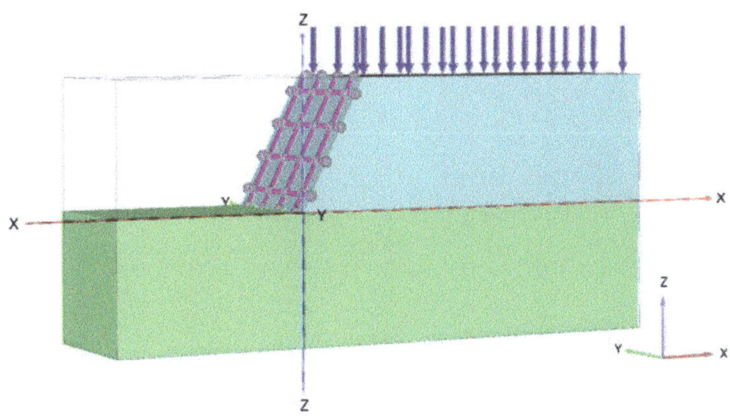

Figure 19. The 1:0.5 model of filled-slope reinforced by a frame with prestressed anchor-plates.

Figure 20 shows the total displacement increment cloud map after the calculation of the model, which can reveal the possible failure mechanism. From Figure 20, it can be clearly seen that the slope failure is an arc sliding failure, and the slip surface will pass over the foot of the slope. Failure may occur in the deformation area of the slope, but the location of the most dangerous slip surface cannot be clearly known. The calculated stability factor is 1.790.

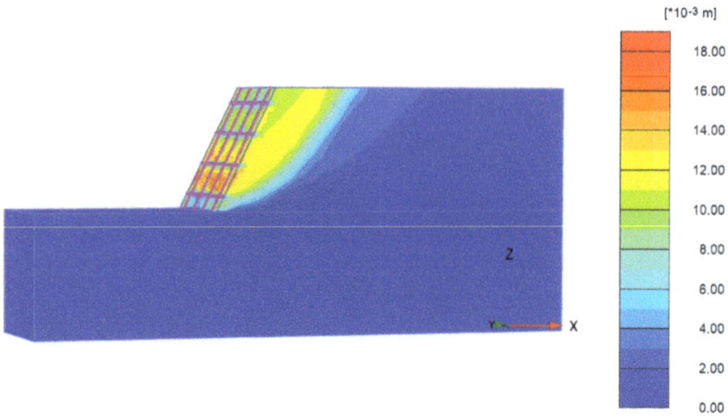

Figure 20. Total displacement increment cloud map of 1:0.5 model.

(3) Ultimate balance method

The slope stability results simulated by PLAXIS 3D are very close to those calculated by this algorithm, but it is not very convincing to verify the rationality of the algorithm in this paper by only relying on the stability results of a slope simulated by a finite element software. Thus, the Bishop method in the SLOPE module of GeoStudio 2012 finite element software is used to calculate the stability of example 1. The anchor-plate is simulated by the "anchor" unit, the specific calculation results are shown in Figure 21. From Figure 21, it can be seen that the most dangerous slip surface passes through the toe of the slope, and the starting point of the slip surface is 6.5 m away from the starting point of the top of slope. The stability factor is 1.840.

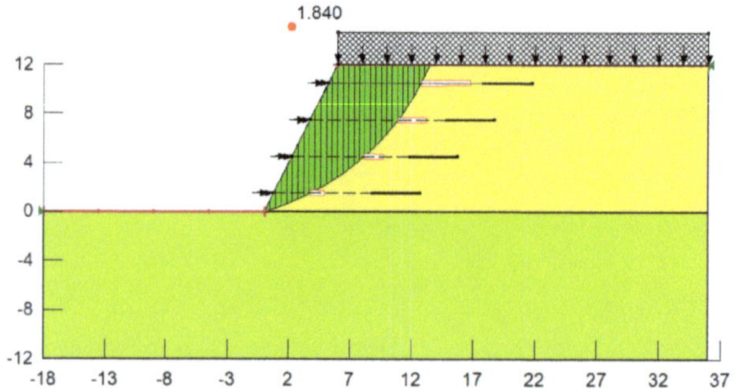

Figure 21. Example 1 calculation results of the stability of Geo.

(4) Comparative analysis

By using the above three methods to calculate the stability of example 1, it is not difficult to see that the slip surfaces obtained by the three methods all pass over the toe of the slope. The stability factor calculated by Geo is the largest, the stability factor calculated by PLAXIS 3D is the second, and the stability factor calculated by this algorithm is the smallest. The stability factor calculated by this method, respectively, differs from the results simulated by Geo and PLAXIS 3D finite element software by 7.1% and 4.6%. By comparing Figure 18 with Figure 21, it can be seen that the position of the most dangerous slip surface obtained by this algorithm is close to that obtained by Geo. Although the location of the most dangerous slip surface cannot be directly obtained in Figure 20, the possible shape of the slip surface can also be seen from the range of areas that may be damaged. The shape of the slip surface reflected in Figure 20 is similar to that of the most dangerous slip surface in Figures 18 and 21, which fully shows that the slip surface search method in this paper is reliable. The stability factor of this algorithm is smaller than that of two finite element software, which also shows that this algorithm has a certain security reserve.

5.2. Example 2

The specific parameters of the filled-slope reinforced by a frame with prestressed anchor-plates in Longnan are as follows. The height of the slope is 12 m, the slope rate is 1: 1, the top load of the slope is $q_0 = 20\ \text{kN/m}^2$, and the soil parameters are shown in Table 4. The horizontal spacing of the anchor-plate is 3 m, and the length, width, and height of the anchor-plate are 3 m, 1 m, and 0.1 m, respectively, and the parameters of anchor-plate for each layer are shown in Table 5.

Table 4. Soil parameters of example 2.

Soil Layer Name	Natural Heavy γ/(kN/m^3)	Cohesive Force c/(kPa)	Angle of Internal Friction φ/(°)	Elastic Modulus E/ kN/m^2)	Friction Coefficient μ
Fill soil	17.5	20	22	30,000	0.35
Foundation soil	21	12	30	45,000	

Table 5. Parameters of anchor-plate of example 2.

Layer Number	Relative Ground Height/m	Length of Steel Tie Rod/m	Prestress Value/kN
Fourth floor	10.5	13.5	50
Third floor	7.5	12.5	80
Second floor	4.5	11.5	110
First floor	1.5	10.5	140

5.2.1. Analysis of Pullout Resistance of Anchor-Plate

The pullout force calculation method of anchor-plates in Zhu [25] and the improved method in this paper are used to calculate the pullout force of anchor-plates in example 2, and the results are shown in Table 6.

Table 6. Calculation results of pullout force of anchor-plates in example 2.

Layer Number	Original Method/kN	The Method of This Paper/kN	Percentage of Difference
Fourth floor	102.89	113.00	9.82%
Third floor	224.08	239.53	6.61%
Second floor	346.47	366.06	5.65%
First floor	408.26	492.59	5.20%

It can be seen from Table 6 that the pullout force of the anchor-plates calculated by this method is larger than that calculated by the method of Zhu [25], and the specific reason has been analyzed in example 1. From Equations (31)–(37), it can be seen that the pullout force of the anchor-plate is not only closely related to its own size, but also related to its position. The thicker the soil layer on the anchor-plate is, the greater the friction resistance of the anchor-plate is, and the greater the pullout resistance is, which is the reason for the greater prestress applied to the anchor-plate near the toe of the slope.

5.2.2. Stability Analysis

Like example 1, the stability calculation method of this paper, strength reduction method and ultimate balance method are used to calculate the stability of example 2, respectively.

(1) The method of this paper

The stability of example 2 is calculated by using the stability analysis process in this paper, and the calculated results are shown in Figure 22. It can be clearly seen from Figure 22 that the most dangerous slip surface is far from the foot of the slope, which belongs to the sliding mode of midpoint circle 1. The position of the most dangerous slip surface is 8 m away from point B at the top of the slope, the end position of the slip surface is 4 m away from point A at the foot of the slope, and the calculated stability factor is 1.612.

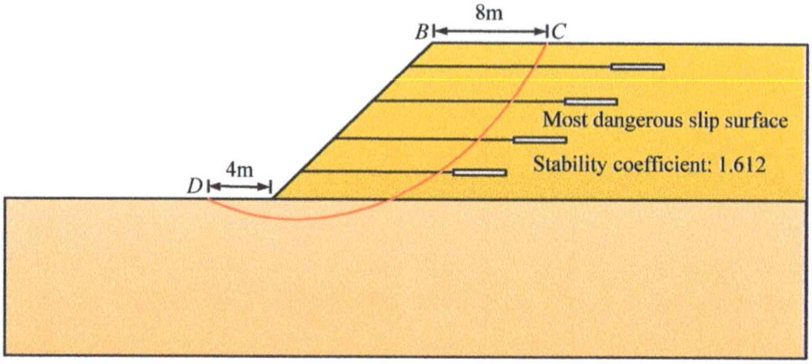

Figure 22. Example 2 calculation results of the stability of this method.

(2) Strength reduction method

The stability of example 2 is calculated with the help of PLAXIS 3D finite element software. The established three-dimensional model of the filled-slope with a slope ratio of 1:1 reinforced by the frame prestressed anchor-plates is shown in Figure 23.

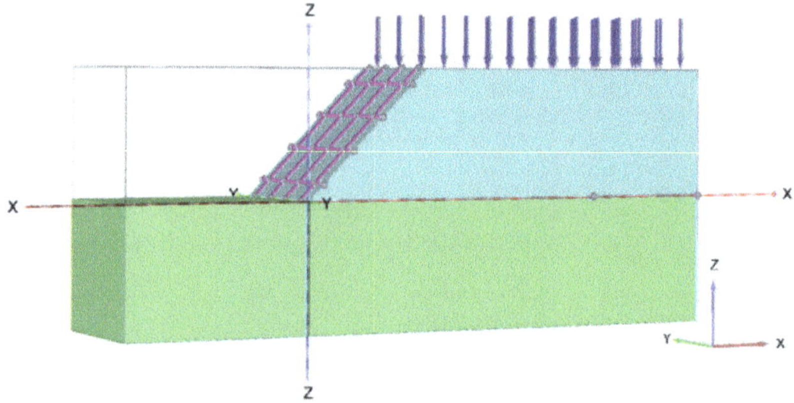

Figure 23. The 1:1 model of filled-slope reinforced by frame with prestressed anchor-plates.

Figure 24 is the total displacement increment cloud map after the calculation of the model, from which it can be clearly see that the failure mode of the slope is circular sliding failure. When the slope is damaged, the slip surface may also be far away from the toe of the slope, except for the case where the slip surface passes through the toe of the slope, and failure may occur in the deformation area of the slope. The calculated stability factor is 1.665.

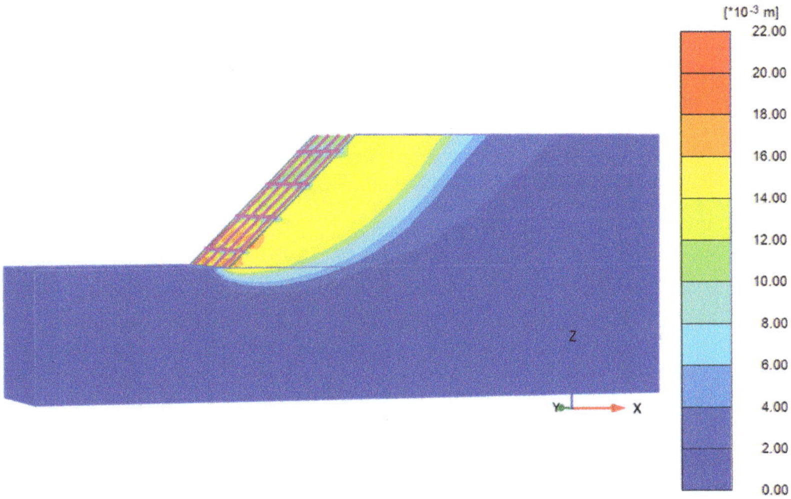

Figure 24. Total displacement increment cloud map of 1:1 model.

(3) Ultimate balance method

The Bishop method in the SLOPE module of GeoStudio 2012 finite element software is used to calculate the stability of example 2, and the specific calculation results are shown in Figure 25. From Figure 18, it can be seen that the most dangerous slip surface is far from the toe of the slope. The starting position of the slip surface is 7.5 m away from the starting point of the top of the slope, the end position of the slip surface is 4 m away from the toe of the slope, and the calculated stability factor is 1.688.

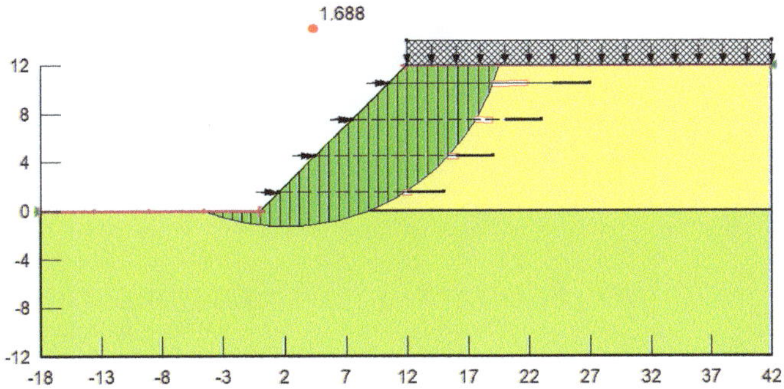

Figure 25. Example 2 calculation results of the stability of Geo.

(4) Comparative analysis

By using the above three methods to calculate the stability of example 2, it can be seen that the stability factor calculated by Geo is the largest, followed by PLAXIS 3D, and the stability factor calculated by this algorithm is the smallest. The stability factor calculated by this method respectively differs from the results simulated by Geo and PLAXIS 3D finite element software by 4.5% and 3.2%. Comparing Figure 22 with Figure 25, the shape of the most dangerous slip surface obtained by this method is similar to that obtained by Geo, both of which are far from the foot of the slope and belong to the sliding mode of midpoint circle 1.

5.3. Summary

Based the analysis of example 1 and example 2, it is concluded that although the slip surface searched by this method is not exactly the same as that searched by Geo software, the form of the slip surface is very similar, and the starting position and end position of the slip surface are similar. The stability factor calculated by this method is smaller than that calculated by the two finite element software, the specific reasons are as follows. When using PLAXIS 3D to calculate the stability factor of filled-slope reinforced by frame with prestressed anchor-plates, in addition to the reinforcement effect of the anchor-plates on the filled-slope, the frame beam also plays a certain role in reinforcing the slope, while the stability calculation method proposed in this paper only considers the effect of the anchor-plates. When using Geo to calculate the stability factor of filled-slope reinforced by a frame with prestressed anchor-plates, the software is a two-dimensional finite element software, and the anchor-plate can only be simulated as an anchor during simulation, which is different from the algorithm in this paper, so the calculated stability factor of the slope is quite different. Although the stability factor of slope calculated by this method is smaller than that calculated by finite element software, it can provide a certain safety reserve for practical engineering when designing a filled-slope reinforced by a frame with prestressed anchor-plates. Therefore, this stability calculation method of a filled-slope reinforced by a frame with prestressed anchor-plates proposed in this paper is reasonable and suitable for the case of arbitrary arc slip surface in the filled-slope reinforced by a frame with prestressed anchor-plates, and it provides some guiding values for the design of practical engineering.

6. Conclusions

In this paper, the calculation formulas of the stability factor under the four arc slip surface of a filled-slope reinforced by a frame with prestressed anchor-plates are derived by using the improved Bishop method, the corresponding search method of the most dangerous slip surface is given, and the calculation formulas of the pullout force of anchor-plates is improved. Based on two examples, the stability results calculated by the proposed algorithm are compared with those calculated by PLAXIS 3D and GeoStudio 2012 finite element software, and the following conclusions are drawn:

(1) Compared with the original calculation method of the pullout force of the anchor-plate, the pullout force of the improved anchor-plate takes into account the friction of the front and rear surface of the anchor-plate and the effect of fill cohesion in the passive earth pressure on the front end of the anchor-plate, which makes the force of the anchor-plate more complete. At the same time, it also makes the calculation theory of pullout force of anchor-plates more applicable.
(2) The stability factor of example 1 calculated by this method differs from the results simulated by PLAXIS 3D and GeoStudio 2012 finite element software by 4.6% and 7.1%, respectively, the stability factor of example 2 calculated by this method differs from the results simulated by PLAXIS3D and GeoStudio 2012 finite element software by 3.2% and 4.5%, respectively, which can meet the engineering requirements.
(3) The stability analysis method of a filled-slope reinforced by a frame with prestressed anchor-plates proposed in this paper is reasonable and suitable for the case of arbitrary arc slip surface in the filled-slope reinforced by a frame with prestressed anchor-plates, and it provides some guiding values for the design of practical engineering.

Author Contributions: Conceptualization: S.Y.; data curation: J.Z.; formal analysis: J.Z., W.L. and S.Y.; investigation: J.Z., W.L. and S.Y.; methodology: J.Z., W.L. and S.Y.; software: J.Z.; writing—original draft: J.Z. and W.L.; writing—review and editing: J.Z., W.L. and S.Y. All authors have read and agreed to the published version of the manuscript.

Funding: This research was funded by the National Natural Science Foundation of China (No. 52168050).

Institutional Review Board Statement: Not applicable.

Informed Consent Statement: The authors listed in this paper all agree to publish this paper.

Data Availability Statement: The data of simulation and theoretical calculation in the article are not freely available due to legal concerns and commercial confidentiality. Nevertheless, all the concepts and procedures are explained in the presented research and parts of the research may be available upon request.

Acknowledgments: The authors of the present work feel grateful and would like to thank Lanzhou University of Technology, GanSu CSCEC Municipal Engineering Investigation And Design Institute Co., Ltd., National Natural Science Foundation of China, for providing financial and technical support in research.

Conflicts of Interest: The authors declare that they have no conflict of interest.

References

1. Luo, X.F.; Xing, G.H.; Zhao, H. Performance of rock-soil slope reinforcement and protection effect in cutting slope construction. *Int. J. Low-Carbon Technol.* **2018**, *13*, 193–197. [CrossRef]
2. Shinoda, M.; Miyata, Y. PSO-based stability analysis of unreinforced and reinforced soil slopes using non-circular slip surface. *Acta Geotech.* **2019**, *14*, 907–919. [CrossRef]
3. Ye, S.H.; Fang, G.W.; Ma, X.R. Reliability analysis of grillage flexible slope supporting structure with anchors considering fuzzy transitional interval and fuzzy randomness of soil parameters. *Arab. J. Sci. Eng.* **2019**, *44*, 8849–8857. [CrossRef]
4. Ye, S.H.; Zhao, Z.F. Allowable displacement of slope supported by frame structure with anchors under earthquake. *Int. J. Geomech.* **2020**, *20*, 04020188. [CrossRef]
5. Nariman, K.; Mahmoud, G. Internal stability analysis of geocell-reinforced slopes subjected to seismic loading based on pseudo-static approach. *Geotext. Geomembr.* **2022**, *50*, 393–407.
6. Keskin, M.S.; Kezer, S. Stability of MSW landfill slopes reinforced with geogrids. *Appl. Sci.* **2022**, *12*, 11866. [CrossRef]
7. Varge, R.; Zlender, B.; Jelusic, P. Multiparametric analysis of a gravity retaining wall. *Appl. Sci.-Basel* **2021**, *11*, 6233. [CrossRef]
8. Varge, R.; Jelusic, P. Failure probability of an optimally designed gravity retaining wall. *ASCE-ASME J. Risk Uncertain. Eng. Syst. Part A Civ. Eng.* **2022**, *8*, 1–11. [CrossRef]
9. Sharma, S.; Saha, A.K.; Lohar, G. Optimization of weight and cost of cantilever retaining wall by a hybrid metaheuristic algorithm. *Eng. Comput. Germany* **2022**, *38*, 2897–2923. [CrossRef]
10. Konai, S.; Sengupta, A.; Deb, K. Seismic behavior of cantilever wall embedded in dry and saturated sand. *Front. Struct. Civ. Eng.* **2020**, *14*, 690–705. [CrossRef]
11. Qiu, Y.Q.; Liu, Y.; Zhang, L.J.; Wang, Z.Q. Influence of lightweight foamed concrete as backfill material on stress and deformation of buttressed earth-retaining wall. *Geofluids* **2021**, *2021*, 1–14. [CrossRef]
12. Zhao, N.Y.; Xu, Y.; Xiang, S.; Song, Y. Pressure distribution lawon vertical plate of cellular-counterfort retaining structure. *Alex. Eng. J.* **2022**, *61*, 3735–3745. [CrossRef]
13. Caputo, G.V.; Conti, R.; Viggiani, G.M.B.; Prum, C. Improved method for the seismic design of anchored steel sheet pile walls. *J. Geotech. Geoenviron.* **2021**, *147*, 1–13. [CrossRef]
14. Gazetas, G.; Garini, E.; Zafeirakos, A. Seismic analysis of tall anchored sheet-pile walls. *Soil Dyn. Earthq. Eng.* **2016**, *91*, 209–221. [CrossRef]
15. *GB 50330-2013*; Technical Code for Building Slop Engineering. China Building Industry Press: Beijing, China, 2013.
16. Ye, S.H.; Fang, G.W.; Chen, C.L.; Zhang, W.L. Numerical and monitoring analyses of new reinforcement technology of frame with prestressed anchor-plates in high filled-slope and its application. *Chin. J. Geotech. Eng.* **2018**, *40*, 57–64.
17. Zhu, Y.P.; Zhu, S.X.; Ye, S.H.; Zheng, X.J.; Zhou, Y.; Wu, Y.Q.; Li, Y.X.; Hui, C.Y.; Li, J.B.; Ma, X.R. Flexible Support System and Construction Method of Prestressed Anchor-plate for Pile-board Wall. China Patent CN10463432A, 20 May 2015.
18. Huang, A.P.; Ye, S.H. Sensitivity of high fill slope stability factors under seismic conditions. *Soil Mech. Found. Eng.* **2020**, *57*, 356–363.
19. Ye, S.H.; Fang, G.W.; Zhu, Y.P. Model Establishment and Response Analysis of Slope Reinforced by Frame with Prestressed Anchors under Seismic Considering the Prestress. *Soil Dyn. Earthq. Eng.* **2019**, *122*, 228–234. [CrossRef]
20. Bai, B.; Zhou, R.; Cai, G.Q.; Hu, W.; Yang, G.C. Coupled thermo-hydro-mechanical mechanism in view of the soil particle rearrangement of granular thermodynamics. *Comput. Geotech.* **2021**, *137*, 104272. [CrossRef]
21. Bai, B.; Yang, G.C.; Li, T.; Yang, G.S. A thermodynamic constitutive model with temperature effect based on particle rearrangement for geomaterials. *Mech. Mater.* **2019**, *139*, 103180. [CrossRef]
22. Li, Z.; Chen, S.Y.; Zhu, Y.P.; Gao, J.X. Calculation method for slope reinforced by grillage per-stressed anchor based on stability of excavation process. *Chin. J. Rock Mech. Eng.* **2014**, *33*, 2964–2970.
23. Zhu, Y.P.; Hou, X.N.; Ma, X.X.; Yang, K.B. Limit analysis of slope stability supported by framed prestressed anchor rods. *Chin. J. Geotech. Eng.* **2021**, *43*, 7–12.
24. He, L.Z.; Yuan, T.F.; Wang, X. Wall stability checking analysis on the suspended-anchor earth retaining wall of two anchor plate. *Highway Eng.* **2012**, *37*, 47–55.

25. Zhu, Y.P.; Tao, J.; Yang, X.H.; Peng, J.G.; Wu, Q. Design and numerical analyses of high-filled-slope strengthened by frame with prestressed anchor-plates. *Rock Soil Mech.* **2020**, *41*, 612–623.
26. Phung, M.V.; Nguyen, D.T.; Doan, L.T.; Nguyen, D.V.; Duong, T.V. Numerical investigation on static bending and free vibration responses of two-layer variable thickness plates with shear connectors. *IJST-T. Mech. Eng.* **2022**, *46*, 1047–1065. [CrossRef]
27. Thom, D.V.; Doan, H.D.; Minh, P.V.; Son, T.N. Finite element modeling for free vibration response of cracked stiffened FGM plates. *Vietnam J. Sci. Technol.* **2020**, *58*, 119–129. [CrossRef]
28. Liu, S.Y. *Soil Mechanics*; China Building Industry Press: Beijing, China, 2020.
29. Zhang, R.H.; Ye, S.H.; Tao, H. Stability analysis of multistage homogeneous loess slopes by improved ultimate balance method. *Rock Soil Mech.* **2021**, *42*, 813–825.
30. Zhu, D.Y.; Deng, J.H.; Tai, J.J. Theoretical verification of rigorous nature of improved Bishop method. *Chin. J. Rock Mech. Eng.* **2007**, *2007*, 455–458.
31. Pereira, T.D.; Robaina, A.D.; Peite, M.X.; Braga, F.D.A.; Rosso, R.B. Performance of analysis methods of slope stability for different geotechnical classes soil on earth dams. *Eng. Agr. Jaboticabal.* **2016**, *36*, 1027–1036. [CrossRef]
32. Sheng, K.; Hong, B.N.; Liu, X.; Shan, H. Modified Bishop method for stability analysis of weakly sloped subgrade under centrifuge model test. *Front. Struct. Civ. Eng.* **2021**, *15*, 727–741. [CrossRef]

applied sciences

MDPI

Article

Large Scale Model Test Study of Foundation Pit Supported by Pile Anchors

Tiantao Su [1,2], Yong Zhou [1,2], Zhengzhen Wang [1,2,*] and Shuaihua Ye [1,2,*]

1 Key Laboratory of Disaster Mitigation in Civil Engineering of Gansu Province, Lanzhou University of Technology, Lanzhou 730050, China
2 Western Center of Disaster Mitigation in Civil Engineering, Ministry of Education, Lanzhou University of Technology, Lanzhou 730050, China
* Correspondence: wangzz@lut.edu.cn (Z.W.); yesh@lut.edu.cn (S.Y.)

Abstract: Due to the special time–space and environmental effects of the foundation pit, there are many unstable factors in the construction process of the field test. The indoor model test can avoid many uncertainties in the construction process due to its operability, which can reduce the interference with the test results and improve the accuracy of the test. In order to further discuss the force-bearing characteristics and deformation laws of loess pits' support structure in Northwest China, a large model test of foundation pit supported by a pile anchor with a geometric similarity ratio of 1:10 was designed and completed. The force and deformation characteristics of the support structure were systematically studied by simulating the conditions of additional load at the pit edge, soil layered excavated, and anchors tensioned. The test results show that: for the pile-anchor support structure, the anchors have significant limiting effects on the displacement of the piles. Especially, when the position of the first row of anchors is closer to the pile top, the displacement of the pile is smaller. The stress state of the piles was changed by the prestressed anchor. The passive stress state of piles is changed from one side of tension and the other side of compression to the active stress state of "S" shape, which makes the distribution of the bending moment of piles more reasonable. The measured earth pressure in the process of soil unloading has a nonlinear distribution, which is different from the classical Rankine earth pressure distribution; specifically, the passive earth pressure in front of the pile is more obvious. In addition, the prestress applied to the anchors has a more significant effect on the internal forces of the other anchors. Compared with sequential tensioning, the prestress loss caused by interval hole tensioning is significantly reduced. The greater the number of spaced holes, the smaller the prestress loss and the better the anchoring effect of the anchor. The results of the study can provide reference for similar model tests, and also for related engineering applications.

Keywords: foundation pit; model test; pile anchor; stress characteristic; deformation

Citation: Su, T.; Zhou, Y.; Wang, Z.; Ye, S. Large Scale Model Test Study of Foundation Pit Supported by Pile Anchors. *Appl. Sci.* **2022**, *12*, 9792. https://doi.org/10.3390/app12199792

Academic Editor: Bing Bai

Received: 24 August 2022
Accepted: 23 September 2022
Published: 28 September 2022

1. Introduction

The limited land available for construction in northwest China has constrained the sustainable development of the region. In order to solve the problem of insufficient land for construction, the development and expansion of underground space has become an effective way to solve the problem, but this has also given rise to many complex foundation pit projects at the same time. [1]. A deep foundation pit is a complex high-risk system, the engineering of which is performed in a multi-phase and multi-field manner, composed of structure, soil and groundwater. It involves the stress path during excavation and unloading of soil, which is closely related to the nature of the soil body [2,3]. However, the current theories and methods of foundation pit design are only based on component design, and the experimental research is also based on field test. Restricted by the factors such as site construction conditions and project attributes, field tests sometimes failed to comprehensively test and analyze the stress characteristics of the support structure.

Therefore, the indoor model test research can make up for the insufficiency of the field test and can qualitatively analyze the mechanical characteristics of the support structure.

There are many scholars who have carried out a lot of research on the model test of foundation pits. Xie Yu [4] studied the bending moment and earth pressure of passive piles subjected to horizontal force through model tests, believing that the bending moment curve of passive piles changes with the distance between the pile and the load boundary and intersects at the lower part of the pile. Xie Jiang [5] simulated the excavation process of the foundation pit through the photoelastic test and studied the force chain and its variation characteristics of the soil around the foundation pit. Liu Yiao [6] conducted a 1:30 indoor model test of foundation pit excavation. He obtained the earth pressure distribution, the horizontal displacement of the wall and the settlement law behind the wall, which were verified by finite element method. Bai bing [7] derived the generalized effective stress principle, which differs from the classical effective stress principle in that the model can automatically take into account the effects of stress paths, temperature paths, and soil structure. Yu Suhui [8] studied the influence of foundation pit excavation and construction on adjacent existing buildings and surrounding soil through model tests; he discussed the horizontal and vertical displacements of the adjacent buildings, the earth pressure around the foundation pit and the variation law of the existing building foundation under different parameters. Xiao Yang [9] used a large model box to simulate the whole process of underground space excavation and studied the influence of several parameters on the h-shaped support system with double-row piles. The test results showed that significant load transfer effects were generated between the rows of piles and that increasing the row spacing within a certain range could result in a more reasonable distribution of bending moments and pile forces. Zheng Gang [10,11] studied the influence of local anchor failure on the soldier pile support system through an indoor model test system. They clarified the mechanism of the continuous failure of the foundation pit and proposed an active, real-time, targeted deformation control method. Tang Deqi [12] studied the evolution of supporting structure performance and earth pressure through large-scale model tests. Wu Honggang [13] studied the synergistic deformation of a combined structure of high-fill slopes through model tests, which was composed of a pile-anchor and reinforced soil. Liang Fayun [14], Lin Hai [15] and Zhao Zhuangfu [16] studied the bearing and deformation characteristics of axially loaded piles and horizontally loaded piles, respectively, through indoor model tests. Zhou Dong [17] measured the deformation of the soil around the passive pile under the action of lateral displacement by model tests and obtained the displacement law of the soil around the pile under different embedment depths. Xia Yuanyou [18,19] studied the ultimate bearing capacity of anchors with different shapes and the failure mechanism of the fixed anchor length. Shen Hong [20] analyzed the horizontal displacement of the pile top, the internal force of the pile body and the single-row cantilever pile group through model tests. Zhou Dequan [21] conducted an in-depth study of the pile-side soil pressure and strain on the pile body of this combined structure of inclined straight piles by model tests, which also included the horizontal displacement of the outer pile. The unilateral force–deformation mechanism and damage mode of the combined structure were revealed. Fan Qiuyan [22] studied the changes in soil stress, anchorage section stress and lateral displacement of support structure in soil foundation pit during excavation and prestressed anchor construction by large indoor model test. Ye Shuaihua [23,24] conducted a systematic study on the deformation of the foundation pit and surrounding structures and analyzed the safety of the adjacent subway tunnel during the excavation of the foundation pit.

As a non-linear elastic material, the stress, strain, and strength properties of soil under stress distortion are quite different from those under conventional perimeter pressure. Conventional model tests are conducted under 1 g (g is the acceleration of gravity), and therefore, the conventional method underestimates the earth pressure and soil deformation caused by the excavation of the foundation pit [25,26]. Centrifugal tests, which can be performed under ng conditions, can create artificial gravity to compensate for this aspect.

Zhang Ga [27] designed a centrifuge model test for foundation excavation used a loading system instead of excavated soil to increase the centrifugal acceleration to achieve a suitable self-weight stress state before excavation. After a series of tests, he was concluded that the real path of stress and deformation of the foundation pit could be reasonably simulated by the centrifuge model test. Sun Yuyong [28] studied the deformation characteristics of the foundation pit through field measurements and centrifuge model tests. It was concluded that the maximum sidewall displacements of the inner and outer pits decreased approximately linearly with the increase of the inner pit spacing but increased with the growth of the inner pit excavation width. Jin Hongliu [29] conducted a series of centrifuge model tests and numerical analyses. The deformation and earth pressure of the retaining structure, and the deformation characteristics of the retaining structure under various restricted soil widths, were studied. Jia Jinqing [30], Ma Xianfeng [31], Xu Qianwei [32], Zhou Qiujuan [33], and Li Lianxiang [34,35] carried out a series of centrifuge model test studies on foundation pits in soft soil areas from different degree perspectives and summarized the internal force and deformation law of the support structure.

Most of the above studies are focused on the economically developed soft soil areas in eastern China. Since the economic level and urban construction in the loess region of northwest China lag behind those in the soft soil region, the research on foundation pit engineering has not kept pace with the soft soil region. Foundation pit engineering has obvious regional characteristics; in other words, there is a huge difference between the engineering characteristics of loess and soft soil, which determines that foundation pits in loess region cannot copy all the experience of foundation pits in soft soil region. For this reason, it is very necessary to carry out the study of indoor model tests for foundation pits in loess areas.

In this paper, based on the previous research [4,8,12,21,22], a large-scale model test study of foundation pits supported by pile-anchors was carried out using a self-researched model box. The displacement and internal force of the supporting pile and the change of the internal force and earth pressure of the anchor during the excavation process are analyzed. The influence of different working conditions such as soil unloading and anchor tensioning on the internal force of the supporting pile is analyzed. The distribution of the axial force and shear stress of the anchor during the pulling process and the influence of the pulling of the adjacent anchor on its internal force are discussed. Eventually, the research results can provide reference for the model test research of pits supported by pile anchors and also for engineering applications.

2. Model Test Design

According to the purpose and conditions of this test, on the basis of a comprehensively considered variety of factors, the test mainly considers 9 key physical dimensions for the test: geometric dimension l, weight γ, displacement δ, strain ε, stress σ, elastic modulus E, Poisson's ratio ν, concentrated force p and additional load q. Based on the similarity theory, the similarity criterion between the model and the prototype is deduced by means of dimension analysis, and then the model design and test are carried out according to the similarity criterion. According to the test requirements and the conditions of the test site, the geometric similarity ratio between the model and the prototype was set as 1:10. The results of the similarity relationship and similarity coefficient of each physical quantity of the model are shown in Table 1.

Table 1. Similarity relationship of model tests.

	Physical Quantity	Relationship Formula	Scale Factor
Material Properties	Strain ε	$C_\varepsilon = 1.0$	1
	Stress σ	$C_\sigma = C_\gamma C_l$	1/10
	Elastic Modulus E	$C_E = 1.0$	1
	Poisson's Ratio ν	$C_\mu = 1.0$	1
	Unit Weight γ	C_γ	1
Geometric Properties	Length l	C_l	1/10
	Displacement δ	$C_\delta = C_l C_\varepsilon$	1/10
Load	Concentrated Force p	$C_p = C_E {C_l}^2$	1/100
	Additional Load q	$C_q = C_E C_l$	1/10

3. Model Test Production

3.1. Model Box and Test Soil

The excavation depth of the indoor scaled-down model test pit is 1.60 m. The support form of the foundation pit is an pile-anchor structure. The self-researched model box's internal dimensions are 3.4 m (length) × 1.4 m (width) × 2.7 m (height), and the model schematic is shown in Figure 1. In order to ensure that the soil deformation can be observed visually during the test, 12 mm-thick plexiglass was used on both sides of the model box. The soil for this test was taken from the site of the foundation pit, and the soil was yellowish brown with a more uniform quality. The soil was filled in layers after sieving, mixing with water and stewing, and each layer was filled with 15 cm, and then leveled and compacted evenly to ensure the compaction coefficient reached 0.90. In order to reduce the influence of the boundary effect on the test results, simethicone was evenly applied inside the model box. Considering the remodeling effect of the soil after the completion of filling, the model was left to stand for more than 24 h before the test was conducted, and samples were taken during the filling process for indoor geotechnical tests. The physical and mechanical parameters of the soil were obtained, as shown in Table 2.

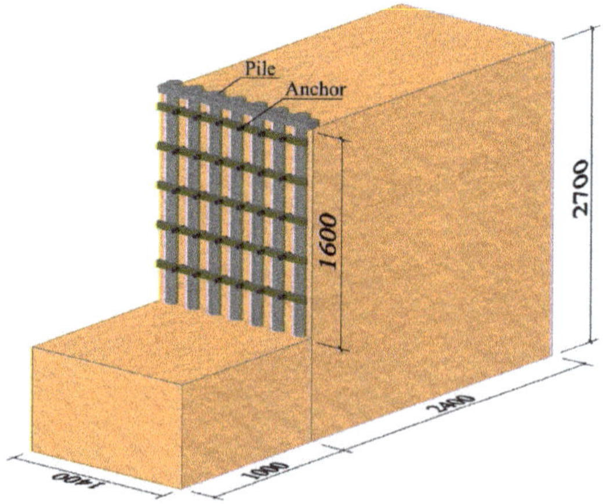

Figure 1. Schematic diagram of the model (mm).

Table 2. Soil parameters.

Soil Layer	Unit Weight/kN/m³	Cohesion/kPa	Angle of Internal Friction/°	Ultimate Bond Strength of Anchor and Soil/kPa
Plain Fill	16.5	16.0	25.0	50.0

3.2. Support Piles and Anchors

The geometric parameters of the piles and anchors are shown in Table 3. The pile body reinforcement adopted No.8 galvanized iron wire. The concrete was made of particulate concrete, with a ratio of 0.49 (water): 1 (cement): 1.370 (sand): 2.486 (stone). Seven piles are arranged along the lateral row, three of the piles are tested, numbered pile 1 to pile 3 in order from left to right. The three piles were tested for strain and pile-top displacement. The anchor body is made of an 8 mm diameter reinforcement; its grouting material ratio is 1 (water): 0.5 (cement). There were 30 anchors in total. The anchors were numbered M*i*-*j* from left to right and from top to bottom, *i* being the row number and *j* being the column number. All anchors in columns 1, 3 and 5 were tested for axial force. The numbering of the components was shown in Figure 2. Considering the fact that the fill was filled horizontally in layers and the anchors were prefabricated in advance, the test ignored the fact that the anchors needed a certain horizontal inclination to ensure the grouting effect in the actual project, so the anchors were arranged horizontally to facilitate the application and measurement of the axial force. The free section of the anchors was covered with corrugated pipe to release the contact between itself and the soil. The anchors were threaded from inside to outside through the holes reserved in the breast beam and fitted with anchors and clips. It was anchored to the breast beam after being tensioned by anchor puller KBT-10T. The anchor length and applied prestress are shown in Table 4.

Table 3. Model parameters.

Project Name	Number of Anchor Rows	Anchorage Body Diameter/mm	Vertical Spacing of Anchors/m	Horizontal Spacing of Anchors/m	Depth of Foundation Pit/m	Diameter of Row Pile/m	Length of Pile/m	Pile Spacing/m
Prototype	5	300	3.5	2.0	17	1.0	25	2.0
Test Model	5	30	0.35	0.2	1.7	0.1	2.5	0.2

Figure 2. Schematic diagram of component numbering.

Table 4. Anchors parameters.

Anchor Position	Total Length/m	Fixed Anchor Length/m	Prestressing/kN
First row	1.9	1.2	3.4
Second row	1.8	1.2	3.4
Third row	1.8	1.2	3.1
Fourth row	1.6	1.1	3.1
Fifth row	1.5	1.0	3.1

3.3. Sensor Arrangement and Loading System

Strain gauges are attached to the surface of the concrete of the support piles and the surface of the anchor's reinforcement. The surface of the strain gauge after pasting was treated with physical protection and waterproof measures. The earth pressure box was fixed on the surface of the pile. The parameters of the sensors are shown in Tables 5 and 6. The location and dimensions of the sensor arrangement are shown in Figure 3. A steel plate is laid on the surface of the soil behind the top of the support pile, and the role of the plate is to transfer the jack load to the soil. The steel plate size is 138 mm (length) * 48 mm (width) * 30 mm (thickness). The additional load around the foundation pit was simulated by a self-designed vertical loading system consisting of a hydraulic jack, a counterforce frame (beam), a load sensor and a pressure-bearing steel plate. The jack axis coincides with the shape center of the steel plate. The equivalent effect of 20 kPa uniform load in this test is 30 kN after conversion.

Table 5. Parameters of strain gauges.

Strain Gauge Type	Measured Material	Resistance Value/Ω	Size/mm	Sensitivity Factor
BE120-3AA-P200	Rebar	119.9 ± 0.1	2.8 × 2.0	2.22 ± 0.1%
BQ120-80AA-P200	Concrete	120.3 ± 0.1	80 × 2.5	2.20 ± 0.1%

Table 6. Parameters of displacement gauges and earth pressure boxes.

Name of the Sensor	Measuring Range	Resolution	Precision
Earth pressure box	0.1 MPa	0.01 kPa	±0.25%F.S.
Displacement gauge	100 mm	-	0.05%F.S.

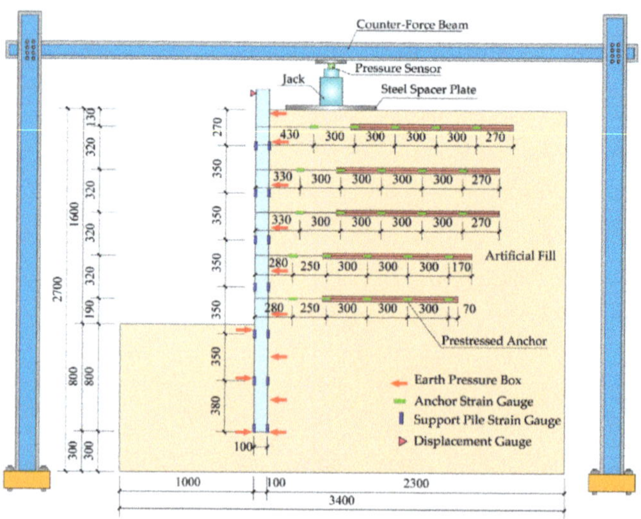

Figure 3. Schematic diagram of test monitoring point arrangement (mm).

3.4. Test Procedure and Data Collection

In order to facilitate the excavation of the soil in layers, six 40 mm thick hardwood boards were used for the front baffle of the model box. A layer of soil was excavated at 24-h intervals, and then the anchors were tensioned and locked. The working conditions design of excavation and anchor tensioning were shown in Table 7. A fine wire mesh sheet was arranged at the pile–soil interface to simulate a shotcrete surface, which is used to avoid soil gushing out from between the piles. The DH3816 N static stress–strain test and analysis system was used for data collection, and the test process and model completion are shown in Figure 4.

Table 7. Test conditions design.

Serial Number	Type of Working Condition	Depth of Each Excavation/mm	Cumulative Excavation Depth/mm	Duration/h	Interval Time from the Previous Working Condition/h
1	Excavate the first layer of soil	180	180	2.0	—
2	Tension the first row of anchor			1.5	2.0
3	Excavate the second layer of soil	320	500	2.0	24.0
4	Tension the second row of anchor			1.5	2.0
5	Excavate the third layer of soil	320	820	2.0	24.0
6	Tension the third row of anchor			1.5	2.0
7	Excavate the fourth layer of soil	320	1140	2.0	24.0
8	Tension the fourth row of anchor			1.5	2.0
9	Excavate the fifth layer of soil	320	1460	2.0	24.0
10	Tension the fifth row of anchor			1.5	2.0
11	Excavation to the bottom of the foundation pit	140	1600	2.0	24.0

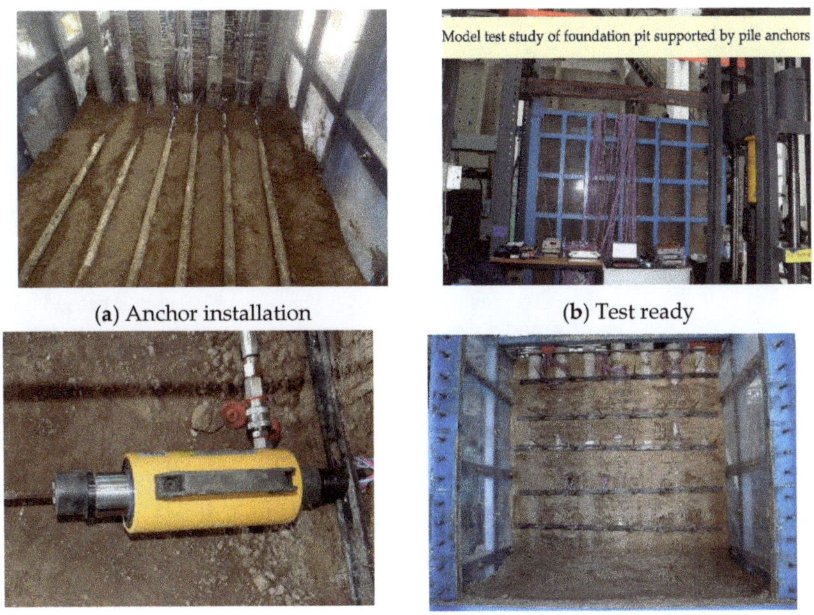

(**a**) Anchor installation (**b**) Test ready

(**c**) Tension anchors (**d**) Complete excavation

Figure 4. Model test process diagram.

4. Analysis of Test Results

4.1. Analysis of Horizontal Displacement of Pile Top

Figure 5 shows the horizontal displacement of the pile top under each working condition. From Figure 5, it can be seen that the displacement of the pile tops tends to increase non-linearly with the excavation of the soil in front of the piles and the tensioning of the anchors. The maximum horizontal displacement of the pile on the edge is 3.78 mm and the middle pile is 3.92 mm. After the prestressed anchors were tensioned, it changed the shear strength of the soil and increased the frictional resistance between the soil and the anchor solid, which slowed down the horizontal displacement of the soil and limited the horizontal displacement of the piles. The control effect of the first anchors on the horizontal displacement of the pile was obvious, which made the piles produce negative displacement and had a slowing effect on the development of the displacement of the subsequent piles. It can be seen that the first anchors played a crucial role in the deformation control of the piles, which further verified that the anchors could effectively control the deformation of the pile and the soil behind the pile.

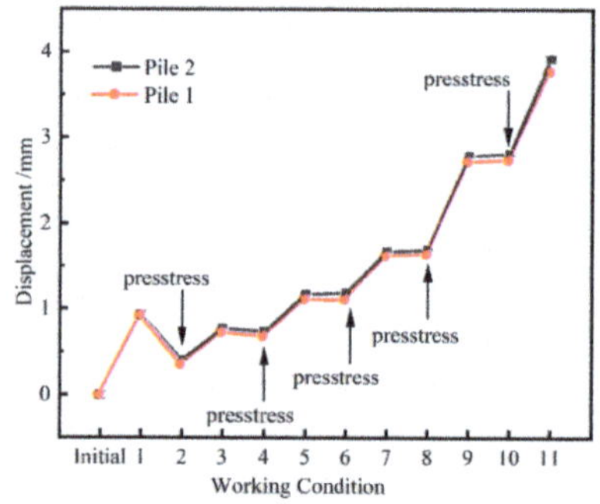

Figure 5. Horizontal displacement of pile top.

Figure 6 shows the horizontal displacement rate of the pile tops under each working condition. It may be explained the fact that the soil resistance in front of the piles was reduced after the soil in front of the piles was excavated, which led to the displacement of the piles in the direction of the airside; after this, the displacement rate showed an increasing trend. When the prestressing anchors were tensioned, the tendency of soil displacement was limited, and the displacement rate showed a decreasing trend. Therefore, each excavation of soil and tensioning of anchor caused the displacement rate to fluctuate up and down. It can be seen that the anchor has obvious effects on limiting the displacement of supporting pile and soil.

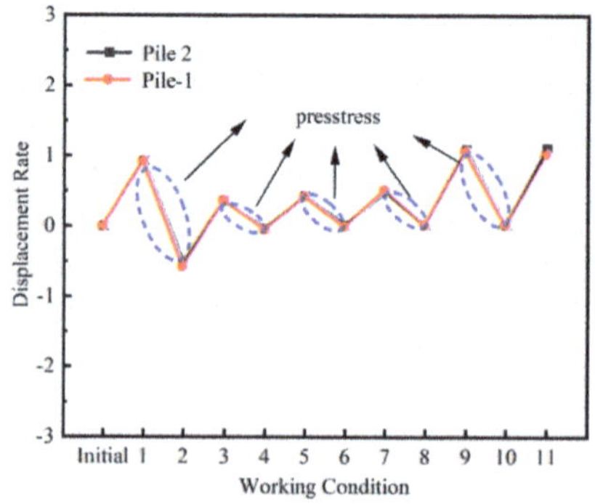

Figure 6. Horizontal displacement rate of pile top.

4.2. Stress Analysis of Supporting Piles

According to the knowledge of material mechanics, the formula for calculating the pile bending moment from strain is [21]:

$$M = EI\frac{\varepsilon_{i1} - \varepsilon_{i2}}{B} \quad (1)$$

where EI is the bending stiffness of the pile body, ε_{i1}, ε_{i2} are the strain values on both sides of the i measuring point, and B is the pile diameter.

It can be seen from the above formula that for the piles, their flexural stiffness EI and pile diameter B are fixed values, so the distribution of the bending moment of the piles' body is the same as the distribution of the strain difference. In order to analyze the change law of the internal force of the pile body intuitively, this paper uses strain instead of bending moment to analyze the strain of pile 2 under different excavation conditions and anchors tensioning conditions, and its distribution is shown in Figure 7. As the soil in front of the pile is excavated, the distribution of strain in the pile body shows itself to be nonlinear. When the pile is in working condition 1 (cantilever stage), the retaining side of the pile is under tension and the maximum value is at the top of the pile, and the strain distribution is in a shape which is small at both ends and large at the middle. As the pile moves from the cantilever state to the single pivot state, the pile bending moment changes from tension on the retaining side to tension on the hollow side. As the supporting pile enters the multi-supported state, the soil resistance in the passive zone gradually increases, the maximum bending moment, the anti-bending point of the pile are shifted downward, and the value of the maximum bending moment on the airside is significantly reduced. Unlike the pile in cantilever state, which is only subject to the soil force, the pile-anchor support structure is subject to the interaction of both soil unloading and anchor tensioning. The bending moment at the embedded end also changes from being characterized by tension on the retaining side in the initial working condition, to tension on the hollow side in the multi-pivot stage, and then returns to a state of tension on the retaining side after excavation is completed. This shows that the anchor is tensioned to form a synergistic effect with the pile and the soil, which effectively changes the force state of the supporting pile and avoids the ultimate force state of the supporting pile.

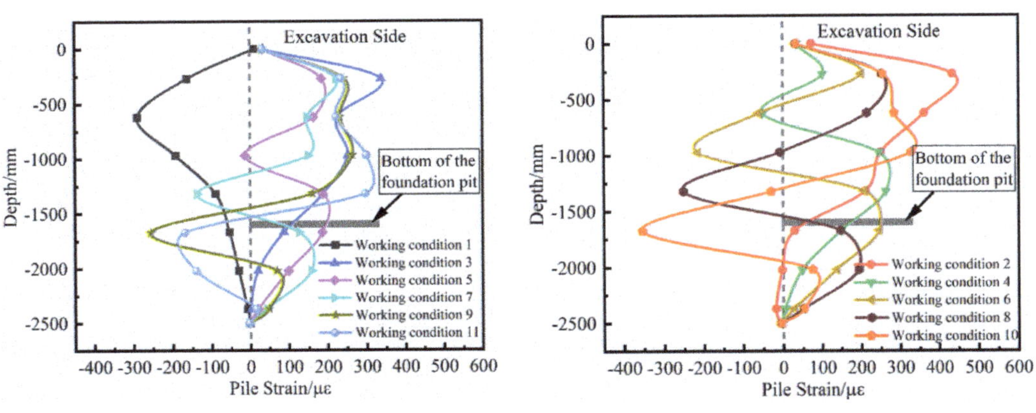

(a) (b)

Figure 7. Strain distribution curve of pile 2. (**a**) Strain distribution of pile 2 under soil excavation conditions; (**b**) strain distribution of pile 2 under anchor tensioning conditions.

Since the spatial position of the supporting pile has an influence on the bending moment of the pile, it is necessary to further analyze the bending moment law of the supporting pile at different positions. The strains of different supporting piles under the final working condition are shown in Figure 8. It can be seen from the figure that the bending moment of pile 2 (intermediate pile) is larger than pile bending moment of pile 1 (side pile). Specifically, the embedded end bending moment is more obviously so. This is due to the boundary effect near the side-pile: the soil near the pile is more self-stabilizing, and the soil pressure acting on the pile is relatively small. This makes the pile bending moment smaller. In the actual situation on site, the supporting piles located at the corner of the pit have smaller pile bending moment than the supporting piles in the middle due to the existence of the shaded angle. The whole pile moment distribution is "S" type, and the reverse bending point is near the bottom of the pit. The supporting pile is under tension on the hollow side above the pit bottom and under tension on the retaining side below the pit bottom, and the maximum positive bending moment is greater than the maximum negative bending moment. The pile–soil–anchor interaction makes the pile moment distribution more reasonable than the cantilever state, reduces the pile internal force and improves the safety and stability of the support structure.

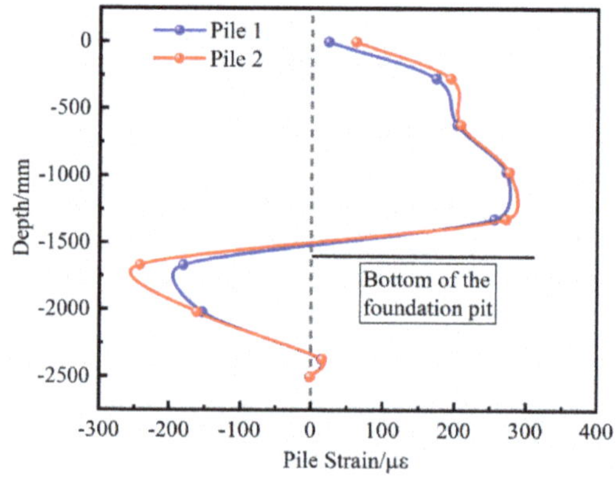

Figure 8. Final strain of different support piles.

4.3. Earth Pressure Analysis

Figure 9 shows the measured soil pressure variation curves under different working conditions. For the measured active earth pressure, the active earth pressure above the bottom of the pit gradually increases under each excavation condition, showing a small distribution at the top and large at the bottom. This is because the supporting pile moves into the pit under the action of the excavated soil in front of the pile and the additional load at the top after the pile, which drives the soil behind the pile to move forward as well, and the soil stress in the active zone increases gradually. For the passive zone soil pressure, as the excavation depth increases, the measured soil pressure below the excavation surface is displaced onto the excavation side due to the supporting pile, and the stress of the soil in front of the pile increases due to the extrusion. In the early stage of excavation, the soil stress in the passive zone in front of the pile increases with depth and shows a small distribution in the upper part and large in the lower part of the figure. At the later stage of excavation, the passive soil pressure in front of the pile is distributed in the shape of large at the top and small at the bottom.

In order to verify the reasonableness of the measured soil pressure, we calculated the Rankine soil pressure of the support structure and compared it with the measured soil pressure. We found that the measured soil pressure is smaller than the Rankine soil pressure and has a non-linear distribution. This is because the Rankine earth pressure theory is based on the rigid retaining wall, which assumes that the soil reaches the ultimate state [36], while the foundation pit does not allow the soil to reach the ultimate state. Therefore, when designing and analyzing the pile-anchored support structure based on the Rankine earth pressure theory of traditional rigid retaining walls, the calculation results may be large.

In the presence of complex excavation conditions or where the deformation range of the support structure is strictly limited, the results calculated by the classical Rankine earth pressure theory may not meet the requirements of the project. Therefore, further research on non-limit states of earth pressure applicable to the foundation pit enclosure structure is needed at a later stage.

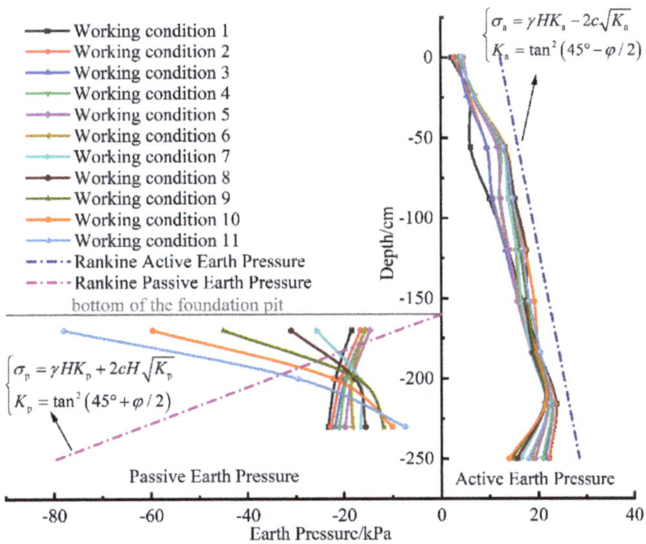

Figure 9. Soil pressure test results under different working conditions compared with Rankine soil pressure.

4.4. Analysis of Internal Forces of Anchors

Figure 10 shows the axial strain distribution of the five anchors at different depths. From the theory of elasticity, it is known that the axial force and axial strain distribution are the same for the same anchors. Therefore, it is more intuitive to use strain instead of axial force when analyzing the axial force. From Figure 10, the axial force of the anchor tends to be zero for the whole length when it is not tensioned. When the prestressing force is applied, the axial force at each point starts to increase and gradually reaches the peak. After the anchor is tensioned, the frictional resistance at the interface between the anchor solid and the soil starts to play, and the anchor section shares part of the axial force coming from the free section. So, the axial force starts to decrease with the length of the anchor solid and drops to the minimum at the end of the anchor [37]. After the anchor tensioning is completed and the jack is removed, the internal force at each measurement point of the anchor shows a certain decrease. At this time, the relative displacement of the anchor solid and its surrounding soil body no longer increases, and the internal force stabilizes. It can also be seen from the figure that the magnitude of the axial force of the anchor also depends on the magnitude of the applied external force. The greater the applied external force, the greater the axial force of the anchor, so the external force within a reasonable range can give full play to the frictional resistance of the anchor solid and the soil. By connecting with the girders and piles, the anchor reduces the deformation of the supporting pile and the soil behind the pile, improves the force state of the pile, and increases the safety and stability of the supporting structure.

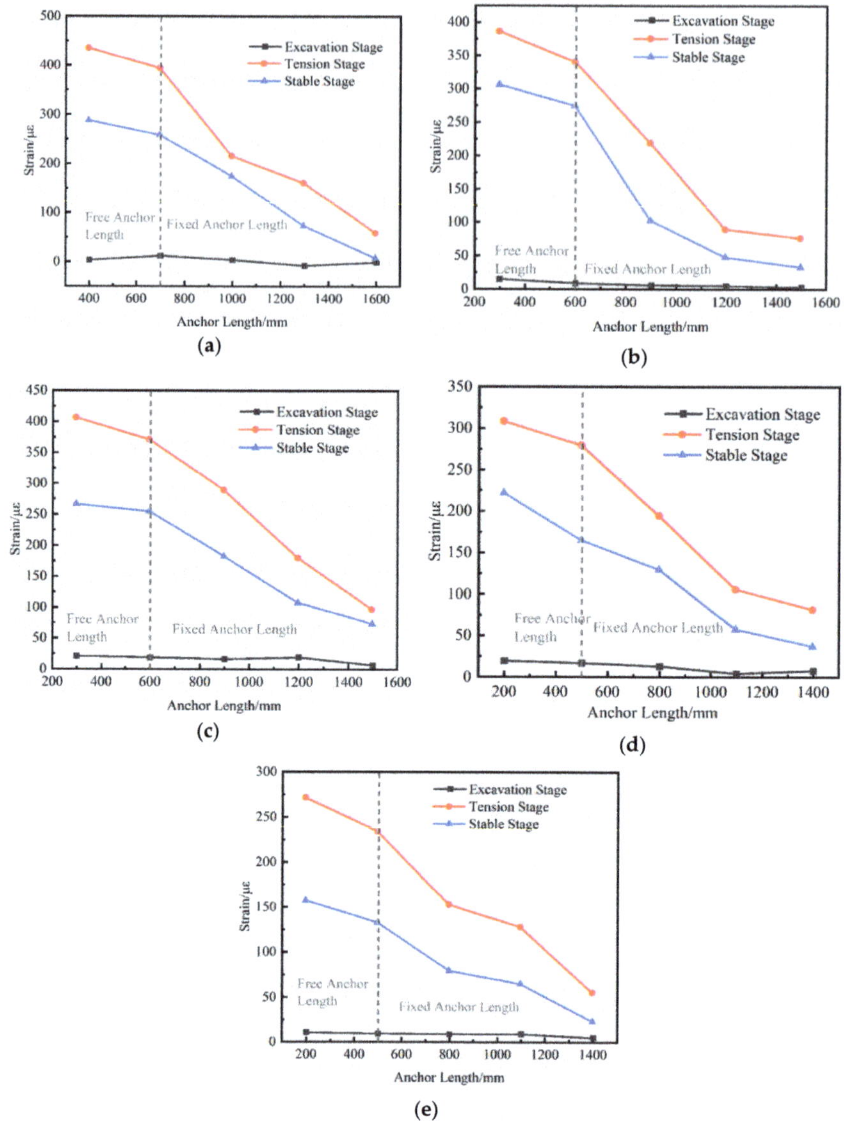

Figure 10. Axial force distribution of anchors. (**a**) Axial force distribution of Anchor M1-3; (**b**) Axial force distribution of Anchor M2-3; (**c**) Axial force distribution of Anchor M3-3; (**d**) Axial force distribution of Anchor M4-3; (**e**) Axial force distribution of Anchor M5-3.

Due to the large number of anchors tested in this experiment, only anchor M2-5 was used to analyze the axial force and shear stress distribution of anchors under different excavation conditions. Figure 11 shows the distribution of the axial force of anchor M2-5 under different excavation conditions. From Figure 11, it can be seen that the anchor axial force starts to increase with the application of prestress and continues to reach its peak after the next layer of soil excavation. The reason for this is that the passive earth pressure starts to decrease due to the continuous unloading of the soil in front of the pile. The pile has a tendency to move into the pit under the action of the soil in the active zone, and this tendency is suppressed by the presence of the anchor. In the later excavation conditions, the anchor shaft force starts to decrease gradually, but it is higher than the initial shaft force after the anchor is tensioned. This indicates that once the anchor is prestressed, its limiting effect on the displacement of the pile and the soil behind the pile will continue to exist, thus ensuring the safety of the support structure.

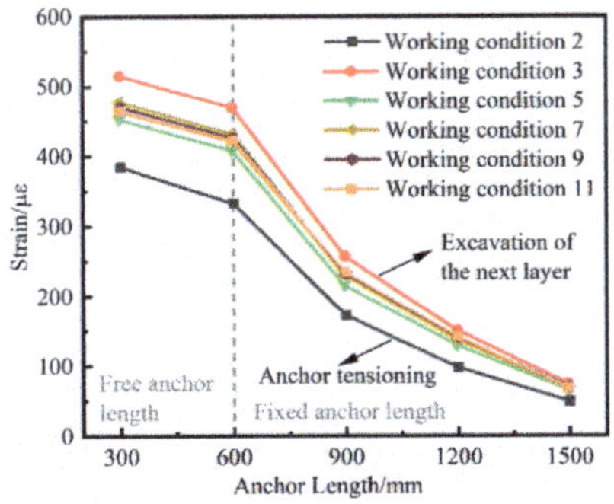

Figure 11. Axial force distribution of M2-5 anchors under excavation conditions.

After the anchor is tensioned, the internal force of any section of the anchored section is equal to the sum of the internal force of the reinforcement and the internal force of the anchor, and the quotient of the difference between the internal force of the two sections and the surface area of the anchor is the shear stress of the anchor. Its calculation formula is given by Liu Yongquan [38].

$$\tau = \frac{(\varepsilon_j - \varepsilon_{j+1})EA_s}{\pi d \Delta s} \tag{2}$$

where E is the modulus of elasticity of the reinforcement, A_s is the cross-sectional area of the reinforcement, ε_i is the strain value of any section i, Δs is the distance between two measurement points and d is the diameter of the anchor solid.

Based on Equation (2), the shear stress distribution of anchor M2-5 under excavation conditions was shown in Figure 12. What can be seen is that, for the tension type anchor, the shear stress distribution of the anchor solid shows a trend of increasing first and then decreasing. It reaches the peak after a certain distance from the beginning of the anchorage section, and then starts to decrease and tends towards zero at the end of the anchorage section. Therefore, for tension anchors, the anchor solid shear stress is mainly borne by the first half of the anchorage section, and the second half of the anchorage section takes up a relatively small proportion. This shows that the anchor solid is not as long as possible, but that there is an optimal anchorage length. After the anchorage section exceeds a certain length, the shear stress provided by the anchor solid is very limited, which is not conducive to the conservation of resources and the sustainable development of the foundation pit project.

Due to the group anchor effect between soil anchors, the tensioning of adjacent anchors affects the internal force of each anchor. The following is an example of anchor M1-3; the effect of tensioning the anchors adjacent to anchor M1-3 on the internal force of M1-3 is investigated, and the results are shown in Figure 13. In order to facilitate visual analysis of the loss of axial force of anchor M1-3, the internal force of anchor M1-3 during tensioning of adjacent anchors is normalized and expressed as the relative value of its axial force. The initial value of the axial force of anchor M1-3 after treatment was 1. The anchor M1-3 was tensioned first, followed by the adjacent anchors in sequence. When the anchors were tensioned according to the sequence in Figure 13a, it can be seen that the axial force of the target anchor M1-3 lost 39% when the anchor M1-2 to the left of M1-3 was tensioned, and the axial force of M1-3 lost another 32% when the anchor M1-4 to its right was tensioned. This was only 29% of its initial value. When tensioned according to the sequence shown in

Figure 13b, the axial force of anchor M1-3 is 74% and 45% of its initial value, respectively. When tensioned in the order shown in Figure 13c, the axial forces of anchors M1-3 are 82% and 62% of their initial values, respectively. It can be seen that the tensioning sequence of anchors has a large effect on the axial force of anchors. When the anchors were tensioned sequentially, the loss of axial force of the target anchors was more obvious. The axial force loss of the target anchor is significantly reduced if tensioning is done at a certain number of intervals. The greater the distance between tensions, the smaller the axial force loss of the target anchor; for the inverse, the greater is the axial force loss. Therefore, the tensioning sequence of anchors in actual construction can follow the principle of a certain distance between tensions in order to reduce the loss of axial force of anchors during the construction process, which leads to the decrease of anchoring effect of anchors. This can effectively avoid the situation that the axial force loss of anchor is too large to control the deformation of the foundation pit and reduce the potential safety of the foundation pit.

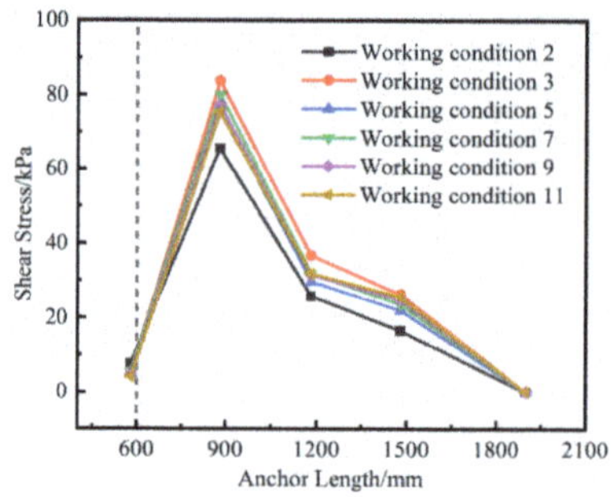

Figure 12. Shear stress distribution of M2-5 anchors under excavation conditions.

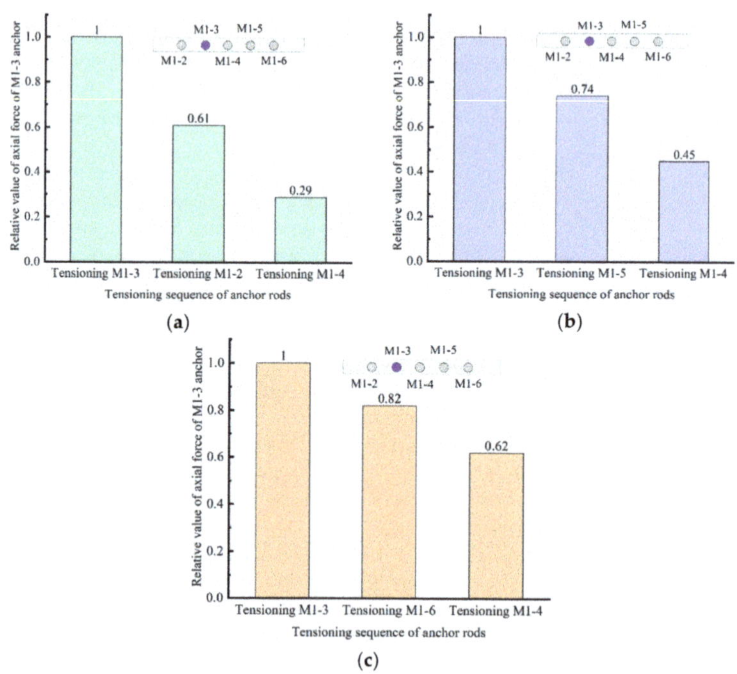

Figure 13. Effect of adjacent anchor tensioned on the axial force of anchor M1-3. (**a**) Sequential tensioning; (**b**) Tensioning at intervals of one hole; (**c**) Tensioning at intervals of two hole.

5. Conclusions

(1) The pile-anchor support structure has good support effect for foundation pit in loess area. The reason is that the interaction of pile–soil–anchor makes the deformation of soil and support structure coordinate with each other, which can effectively restrain the lateral deformation of soil after the pile, especially as the position of the first anchor point has a more obvious limiting effect on the displacement of support pile.

(2) Compared with cantilever piles, where the bending moment is distributed in tension on one side and compression on the other side, the pile-anchor support structure has an "S"-type distribution of bending moment due to the presence of prestressed anchors. With the excavation of the soil and the effect of the prestressing anchor, the maximum value of the pile bending moment and the position of the anti-bending point are decreasing. This avoids the possible extreme force state of the supporting pile, thus improving the bearing capacity of the supporting pile.

(3) Different from the classical linear distribution of earth pressure, the active and passive earth pressure of the pile-anchor-retaining structure show nonlinear distribution due to the unloading effect caused by soil excavation on the passive side and the action of prestressed anchor. The distribution mode of the measured earth pressure is quite different from that of Rankine earth pressure, and the measured earth pressure is smaller than that of Rankine earth pressure. Therefore, the use of earth pressure theory in the design and analysis of foundation pit support should be carefully considered and verified with engineering experience, so as to ensure that the design of foundation pit support is reasonable, economic and safe.

(4) After the prestressing force is applied to the anchors, the axial force starts to decrease at the beginning of the anchorage section and decreases to the minimum at the end of the anchorage section. The shear stress of the anchors starts to increase at the beginning of the anchorage section, reaches the peak after a certain distance, and then starts to decrease gradually.

(5) Due to the effect of group anchors, the tensioning of adjacent anchors will reduce each other's internal force and affect the anchoring effect. It is recommended that anchor rods at the same horizontal position be tensioned at interval holes. This can reduce the prestress loss of the front anchors, and the larger the interval distance, the smaller the prestress loss and the better the corresponding anchorage effect.

Author Contributions: Conceptualization, Y.Z. and T.S.; methodology, T.S. and Z.W.; formal analysis, T.S., Z.W., Y.Z. and S.Y.; investigation, T.S.; data curation, T.S.; writing—original draft preparation, T.S.; writing—review and editing, Y.Z., Z.W. and S.Y. All authors have read and agreed to the published version of the manuscript.

Funding: This research was funded by the National Natural Science Foundation of China (Grant No. 51568042) and Technology Project of Gansu Provincial Department of Housing and Urban-Rural Development Construction (Grant No. JK2022-03).

Institutional Review Board Statement: Not applicable.

Informed Consent Statement: Not applicable.

Data Availability Statement: The data of simulation and monitoring in the article is not freely available due to legal concerns and commercial confidentiality. Nevertheless, all the concepts and procedures are explained in the presented research and parts of the research may be available upon request.

Acknowledgments: The authors of the present work feel grateful and would like to thank National Natural Science Foundation of China, Western Center of Disaster Mitigation in Civil Engineering.

Conflicts of Interest: The authors declare that they have no conflict of interest.

References

1. Zheng, G. Method and application of deformation control of excavations in soft ground. *Chin. J. Geotech. Eng.* **2022**, *44*, 1–36+201. [CrossRef]
2. Bai, B.; Wang, Y.; Rao, D.Y.; Bai, F. The effective thermal conductivity of unsaturated porous media deduced by pore-scale SPH simulation. *Front. Earth Sci.* **2022**, *10*, 943853. [CrossRef]
3. Bai, B.; Yang, G.C.; Li, T.; Yang, G.S. A thermodynamic constitutive model with temperature effect based on particle rearrangement for geomaterials. *Mech. Mater.* **2019**, *139*, 103180. [CrossRef]
4. Xie, Y.; Zhang, S.H.; Zhou, D.Q. Experimental Study of Mechanical Behavior of Passive Loaded Piles Adjacent to Piled Foundation. *Ksce J. Civ. Eng.* **2018**, *22*, 3818–3826. [CrossRef]
5. Xie, J.; Lu, X.J.; Wang, J.N.; Yang, L.; Han, X.G. Force Chain Effect of Deep Foundation Pit Supported by Soldier Piles in the Sand-Gravel Layer. *Shock. Vib.* **2021**, *2021*, 9617807. [CrossRef]
6. Liu, Y.A.; Wang, C.M.; Gao, R.Y.; Li, B.L.; Liu, X.Y.; Liang, Z.; Jiang, N.; Chen, Z.D. Experimental Study for the Embedded Depth of Support Structure Foundation Pit in Granite Residual Soil Area. *Adv. Civ. Eng.* **2020**, *2020*, 6645942. [CrossRef]
7. Bai, B.; Zhou, R.; Cai, G.Q.; Hu, W.; Yang, G.C. Coupled thermo-hydro-mechanical mechanism in view of the soil particle rearrangement of granular thermodynamics. *Comput. Geotech.* **2021**, *137*, 104272. [CrossRef]
8. Yu, S.H.; Geng, Y.C. Influence Analysis of Underground Excavation on the Adjacent Buildings and Surrounding Soil Based on Scale Model Test. *Adv. Civ. Eng.* **2019**, *2019*, 6527175. [CrossRef]
9. Xiao, Y.; Wang, X.G.; Yu, F.; Wang, Z.J. Experimental Investigation of h-Type Supporting System for Excavation beneath Existing Underground Space. *Buildings* **2022**, *12*, 635. [CrossRef]
10. Zheng, G.; Lei, Y.W.; Cheng, X.S.; Li, X.Y.; Wang, R.Z. Influences and mechanisms of anchor failure on anchored pile retaining system of deep excavations. *Chin. J. Geotech. Eng.* **2020**, *42*, 421–429. [CrossRef]
11. Zheng, G.; Wang, R.Z.; Cheng, X.S.; Lei, Y.W.; Li, X.Y.; Zhou, Q. Mechanism and control of progressive collapse caused by local anchor failure in multiple-level anchored pile excavation. *Chin. J. Geotech. Eng.* **2022**. Available online: http://kns.cnki.net/kcms/detail/32.1124.TU.20220512.1844.006.html (accessed on 13 May 2022).
12. Tang, D.Q.; Yu, F.; Chen, Y.T.; Liu, N.W. Model excavation tests on double layered retaining structure composed of existing and supplementary soldier piles. *Rock Soil Mech.* **2019**, *40*, 1039–1048. [CrossRef]
13. Wu, H.G.; Pai, L.F.; Lai, T.T.; Zhang, J.D.; Zhang, Y.P.; Li, Y.R. Study on cooperative performance of pile-anchor-reinforced soil combined retaining structure of high fill slopes in mountainous airports. *Chin. J. Rock Mech. Eng.* **2019**, *38*, 1498–1511. [CrossRef]
14. Liang, F.Y.; Yao, G.S.; Chen, H.B.; Li, J.P. Model tests on behavior of axially loaded piles subjected to lateral soil movement. *Chin. J. Geotech. Eng.* **2010**, *32*, 1603–1609.
15. Lin, H.; Lei, G.H.; Xu, L.; Lei, G.G. Similarity analysis of deflection behavior of laterally loaded piles in 1g model tests. *J. Cent. South Univ.* **2012**, *43*, 3639–3645.
16. Zhao, Z.F.; YE, S.H.; Zhu, Y.P.; Tao, H.; Chen, C.L. Scale model test study on negative skin friction of piles considering the collapsibility of loess. *Acta Geotech.* **2021**, *17*, 601–611. [CrossRef]
17. Zhou, D.; Liu, H.L.; Zhang, W.G.; Ding, X.M.; Yang, C.Y. Transparent soil model test on the displacement field of soil around single passive pile. *Rock Soil Mech.* **2019**, *40*, 2686–2694. [CrossRef]
18. Xia, Y.Y.; Chen, C.; Ni, Q. Pull-out mechanism of continuous ball shape anchors in transparent soil. *Chin. J. Geotech. Eng.* **2017**, *39*, 804–812. [CrossRef]
19. Xia, Y.Y.; Chen, C.; Ni, Q. Comparative modelling of pull-out process of four different anchorages by using transparent soil. *Chin. J. Geotech. Eng.* **2017**, *39*, 399–407. [CrossRef]
20. Shen, H.; Li, X.; Lei, M.Q.; Xu, W.B.; Yu, X.L. Conception and model test of shear bond supporting system. *Rock Soil Mech.* **2019**, *40*, 2574–2580. [CrossRef]
21. Zhou, D.Q.; Feng, C.X.; Xiao, C.; Hu, P.; Cao, Z.Y.; Liu, C.X.; Zhu, Q.; Cai, G.Q. Experimental study on the failure mode of inclined-straight pile composite structure on declining soft foundation under single side load. *China J. Highw. Transp.* **2021**, *34*, 201–214. [CrossRef]
22. Fan, Q.Y.; Chen, B.; Shen, B. Model test research of bolt supporting of foundation pit considering construction process. *Rock Soil Mech.* **2005**, *26*, 1874–1878. [CrossRef]
23. YE, S.H.; Zhao, Z.F.; Wang, D.Q. Deformation analysis and safety assessment of existing metro tunnels affected by excavation of a foundation pit. *Undergr. Space* **2021**, *6*, 421–431. [CrossRef]
24. Ye, S.H.; Ding, S.H.; Gong, X.N.; Gao, S.; Chen, C.L. Monitoring and numerical simulation of deep foundation pit of a subway station in Lanzhou. *Chin. J. Geotech. Eng.* **2018**, *40*, 177–182. [CrossRef]
25. Chen, Y.M.; Han, C.; Ling, D.S.; Kong, L.G.; Zhou, Y.G. Development of geotechnical centrifuge ZJU400 and performance assessment of its shaking table system. *Chin. J. Geotech. Eng.* **2011**, *33*, 1887–1894.
26. Ma, X.F.; Cao, M.Y. Centrifuge model tests on excavation in Shanghai soft soil layers using in-flight excavation tools. *China Civ. Eng. J.* **2022**. [CrossRef]
27. Zhang, G.; Yan, G.C. In-Flight Simulation of the Excavation of Foundation Pit in Centrifuge Model Tests. *Geotech. Test. J.* **2016**, *39*, 59–68. [CrossRef]
28. Sun, Y.Y.; Xiao, H.J. Wall Displacement and Ground-Surface Settlement Caused by Pit-in-Pit Foundation Pit in Soft Clays. *Ksce J. Civ. Eng.* **2021**, *25*, 1262–1275. [CrossRef]

29. Jin, H.L.; Zhang, G.; Yang, Y.S. Experimental and numerical study on behavior of retaining structure with limited soil. *Geomech. Eng.* **2021**, *26*, 77–88. [CrossRef]
30. Jia, J.Q.; Gao, J.C.; Tu, B.X.; Zhang, L.; Wang, H.T.; Gao, R.Z. Centrifugal model test of flexible retaining structures with pressured prestressed anchor in deep excavation. *Rock Soil Mech.* **2017**, *38*, 304–310. [CrossRef]
31. Ma, X.F.; Zhang, H.H.; Zhu, W.J.; Zheng, Y.F.; Xu, Q.W. Centrifuge model tests on deformation of ultra-deep foundation pits in soft ground. *Chin. J. Geotech. Eng.* **2009**, *31*, 1371–1377.
32. Xu, Q.W.; Ma, X.F.; Zhu, H.H.; Ding, W.Q. Centrifugal model test on extra-deep foundation pit excavations in soft ground. *China Civ. Eng. J.* **2009**, *42*, 154–161. [CrossRef]
33. Zhou, Q.J.; Chen, X.P.; Xu, G.M. Centrifugal model test and numerical simulation of soft soil foundation pit. *Chin. J. Rock Mech. Eng.* **2013**, *32*, 2342–2348.
34. Li, L.X.; Fu, Q.H.; Huang, J.J. Centrifuge model tests on cantilever foundation pit engineering in sand ground and silty clay ground. *Rock Soil Mech.* **2018**, *39*, 529–536. [CrossRef]
35. Li, L.X.; Fu, Q.H.; Huang, J.J. Research and application of a new excavation method in centrifuge model tests on foundation pit engineering. *Chin. J. Rock Mech. Eng.* **2016**, *35*, 856–864. [CrossRef]
36. Dang, F.N.; Zhang, L.E.; Wang, X.; Ding, J.L.; Gao, J. Analysis of earth pressure on retaining walls with limited displacement based on elastic theory. *Chin. J. Rock Mech. Eng.* **2020**, *39*, 2094–2103. [CrossRef]
37. Zheng, X.G.; Zhang, N.; Xue, F. Study on stress distribution law in anchoring section of prestressed bolt. *J. Min. Saf. Eng.* **2012**, *29*, 365–370.
38. Liu, Y.Y.; Liu, X.R.; Yang, Z.P.; Kang, J.W. Field test on anchorage performance of different types of prestressed cables. *Chin. J. Rock Mech. Eng.* **2016**, *35*, 275–283. [CrossRef]

applied sciences

MDPI

Article

Calculation for Permanent Displacement of Single Slip Surface of Multi-Stage Loess Slope Based on Energy Method

Xiaobing Zhang [1,2], Xin Zhang [1] and Shuaihua Ye [3,*]

1 School of Civil Engineering, Shandong Jianzhu University, Jinan 250101, China
2 Northwest Research Institute Co., Ltd. of C.R.E.C, Lanzhou 730030, China
3 School of Civil Engineering, Lanzhou University of Technology, Lanzhou 730050, China
* Correspondence: yeshuaihua@163.com

Abstract: At present, there are still some gaps in the theoretical calculation of the permanent displacement of the multi-stage loess slope under the action of earthquake. Therefore, this paper firstly uses the GEO-Studio finite element software to explore the influence of the comprehensive slope rate on the permanent displacement when the slope rate of each grade of multi-stage loess slope changes and the stage of multi-stage slope changes. The results show that it is feasible to use the comprehensive slope rate to calculate the permanent displacement of the sliding body of the multi-stage loess slope under the action of earthquake. On the basis of this conclusion, in order to simplify the calculation of permanent displacement of potential sliding soil, the other geometric parameters of the multi-stage loess slope are replaced by the comprehensive slope rate, combined with the Newmark slider displacement analysis method and energy conservation principle, and then the calculation method of permanent displacement of potential sliding soil of multi-stage loess slope under the action of earthquake is deduced. Through an example, the permanent displacement calculated by the proposed algorithm are compared with those calculated by the PLAXIS 3D software and GEO-Studio software. The results show that the permanent displacement calculated by this method is close to that calculated by the GEO-Studio software, and the difference is only 2%, and thus, the rationality of the proposed method for calculating the permanent displacement of the multi-stage loess slope under the action of earthquake is verified. The algorithm proposed in this paper provides a theoretical reference for the calculation of the permanent displacement of multi-stage loess slope under the action of earthquake.

Keywords: energy method; multi-stage loess slope; single slip surface; permanent displacement; calculation method

Citation: Zhang, X.; Zhang, X.; Ye, S. Calculation for Permanent Displacement of Single Slip Surface of Multi-Stage Loess Slope Based on Energy Method. *Appl. Sci.* **2022**, *12*, 8426. https://doi.org/10.3390/app12178426

Academic Editor: Bing Bai

Received: 29 July 2022
Accepted: 19 August 2022
Published: 24 August 2022

1. Introduction

Loess is a kind of unsaturated soil with large pores, and loess in China is mainly distributed in the northwest region [1–5]. In recent years, as the "Western Development" strategy and "Belt and Road Initiative" strategy have entered a key stage of accelerated development, there will be a large number of multi-stage loess slopes when various infrastructure is built along them. However, China is also a country with frequent earthquakes, and it is very easy to cause multi-stage loess slope landslides under the action of earthquake, which seriously threatens people's lives and property safety [6,7]. Therefore, it is particularly important to study the calculation of permanent displacement of multi-stage loess slope under the action of earthquake [8–10]. At present, the research on the calculation of permanent displacement of the slope under the action of earthquake is as follows. The finite slider displacement analysis method was put forward by Newmark [11] in 1965. It is suggested that the index of possible displacement of potential sliding soil in earth–rockfill dam during earthquake should be used in evaluating the seismic safety of earth–rockfill dam, and a theoretical model for calculating the permanent displacement of potential sliding soil is put forward. Now, this method is widely used in the evaluation of slope stability during earthquake. Since the Newmark slider displacement calculation model was proposed, many scholars have proposed many modified models based on this theory to consider the dynamic response of sliding blocks. These methods include the slider analysis method of coupling and decoupling, among which the simplified model proposed by Makdisi et al. [12] is the most representative one. Steven et al. [13] introduced the dynamic response of the material on the potential sliding surface into the traditional calculation method of slider displacement. Ellen et al. [14–16] compared the difference between decoupling and coupling methods in calculating permanent displacement, proposing a coupling-based analysis model considering sliding mass and sliding plane nonlinearity. Zhang et al. [17,18] provide a strict method for the general calculation of elastodynamic plane problems according to the radial wave function expansion and transfer matrix method. Many scholars have also developed a variety of empirical prediction models for the permanent displacement of potential slip surface of the slope under the action of earthquake, where the permanent displacement is a function of slope geometry, soil parameters or one or more ground motion parameters. Bray et al. [19] proposed a simplified semi-empirical program, which can be used to evaluate the displacement of natural soil or landfill structures that may slip during earthquake. Hsieh et al. [20] used the regression model and regression method to improve the JB93 formula and JB98 formula to calculate Newmark displacement by discussing the relationship between Avias strength and Newmark displacement and the relationship between critical acceleration and Newmark displacement. Du et al. [21] quantitatively studied the influence of the change in slope properties on calculating the displacement of the slope in a Newmark rigid body and fully coupled analysis. Through Monte Carlo simulation, the changes in soil strength, groundwater level, nonlinear soil properties and other parameters are studied, and the influence of these parameters on the displacement prediction of rigid and soft slope is studied comprehensively. Some scholars have also extended the original model with only one potential sliding surface and slider to a nested Newmark slider displacement calculation model with multiple potential sliding surfaces and sliders. Song et al. [22] derived the sliding model of two blocks on the sliding model of one block, and then derived the three-block model, discussed seven possible sliding modes, and put forward a method for calculating the permanent seismic displacement of infinite slope. In addition to the above research, some scholars have also studied the displacement of the slope strengthened by supporting structure under the action of earthquake. Considering the reinforcement effect of a frame with a prestressed anchor on the slope, Ye et al. [23–26] established the analysis model of dynamic response and displacement of the slope under the action of earthquake, deduced its calculation theory, and put forward the analysis method of dynamic response and displacement of the slope strengthened by a frame with prestressed anchor under the action of earthquake.

To sum up, it is not difficult to see that the existing research on the permanent displacement of slope under earthquake is mainly focused on the correction of the Newmark slider displacement calculation model, the development of empirical prediction models of permanent displacement, the establishment of the nested Newmark slider displacement calculation model with multiple slip surfaces and sliders, and the calculation for permanent displacement of single-stage slope strengthened by supporting structure. However, there are still some gaps in the theoretical calculation of the permanent displacement of multi-stage slope and multi-stage loess slope. At present, there are still some gaps in the theoretical calculation of the permanent displacement of multi-stage loess slope under the action of earthquake. Therefore, this paper firstly uses the GEO-Studio finite element software to explore the influence of the comprehensive slope rate on the permanent displacement when the slope rate of each grade of multi-stage loess slope changes and the stage of multi-stage slope changes. The results show that it is feasible to use the comprehensive slope rate to calculate the permanent displacement of the sliding body of multi-stage loess slope under the action of earthquake. On the basis of this conclusion, in order to simplify the calculation of permanent displacement of potential sliding soil, the other geometric parameters of multi-stage loess slope are replaced by the comprehensive slope rate, combined with Newmark slider displacement analysis method and energy conservation principle, and then the calculation method of permanent displacement of potential sliding soil of multi-stage loess slope under the action of earthquake is deduced. Through an example, the permanent displacement calculated by the proposed algorithm are compared with those calculated by PLAXIS 3D finite element software and GEO-Studio finite element software, and the rationality of the proposed method for calculating the permanent displacement of multi-stage loess slope under the action of earthquake is verified. The algorithm proposed in this paper provides a theoretical reference for the calculation of permanent displacement of multi-stage loess slope under the action of earthquake.

2. Influence of the Comprehensive Slope Rate on Permanent Displacement

In the process of excavation of multi-stage loess slope, the slope is unloaded by setting the width of the platform and changing the original slope rate, which leads to the redistribution of stress in the slope. In terms of geometric characteristics, compared with single-stage loess slope, multi-stage loess slope has more definitions of unloading platform width, each grade height, each grade slope rate and slope series. The comprehensive slope rate is the ratio of the total slope height to the distance from the top to the toe of slope of the multi-stage slope, so the comprehensive slope rate can reflect these geometric parameters quantitatively to a certain extent. First of all, in this paper, the influence of comprehensive slope on the permanent displacement of multi-stage loess slope under the action of earthquake is explored by numerical simulation software, and then the well-defined physical quantity of comprehensive slope is introduced to simplify the process of theoretical calculation. In terms of qualitative permanent displacement, a minimum safety factor F_s less than 1.2 is defined as the failure of the slope, and the corresponding horizontal displacement of the sliding soil is the permanent displacement of the slope under the action of earthquake, and the subsequent permanent displacement is expressed according to this definition. Taking the working condition A-1 in Table 1 as an example, the SLOPE/W module based on quasi-static method is used to calculate the stability of the slope after earthquake in GEO-Studio numerical simulation software, and the slip surface form and minimum safety factor of the slope are shown in Figure 1.

Table 1. Each grade of multi-stage loess slope changes the calculation condition.

Working Condition	Slope Rate of Each Grade			Working Condition	Slope Rate of Each Grade		
A-1	1:1	1:1	1:1	A-7	1:0.85	1:0.9	1:1.25
A-2	1:0.95	1:1	1:1.05	A-8	1:0.8	1:1.1	1:1.1
A-3	1:0.95	1:0.95	1:1.1	A-9	1:0.8	1:0.95	1:1.25
A-4	1:0.9	1:1.05	1:1.05	A-10	1:0.75	1:1.1	1:1.15
A-5	1:0.9	1:0.9	1:1.2	A-11	1:0.75	1:1.05	1:1.2
A-6	1:0.85	1:1.05	1:1.1	A-12	1:0.75	1:1	1:1.25

It can be seen from Figure 1 that the minimum safety factor is 0.903. According to the definition of this paper, the slope has been destroyed and has produced permanent displacement. By calculating the rest of the working conditions in this paper, it is found that the minimum safety factor is less than 1.2, which means that these slopes all cause permanent displacement.

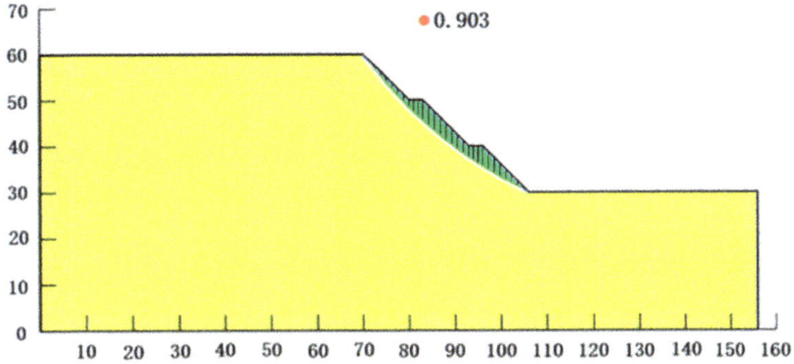

Figure 1. Result of seismic stability analysis of slope under A-1 working condition.

2.1. When the Slope Rate of Each Grade of Multi-Stage Slope Is Changed

In order to explore the influence of the change in the slope rate of each grade on the permanent displacement of potential sliding soil of multi-stage loess slope under the action of earthquake, a three-stage loess slope model is established by using GEO-Studio geotechnical finite element software. The total height of the slope is H =30 m, the height of each slope is $h_1 = h_2 = h_3 = 10$ m, the width of the platform is 3 m, the friction angle of soil is $\varphi = 15°$, the unit weight of soil is $\gamma = 16.6$ kN/m^3, the slope rate of each grade is 1:0.75~1:1.25, and the slope as a whole shows a steep trend at the upper part and gentle at the lower part. Additionally, a total of 12 working conditions are set for simulation calculation when the comprehensive slope rate is 1:1.2; the specific working conditions are shown in Table 1.

The A-1 analysis model is shown in Figure 2. The model soil parameters are loess-like silt, the linear elastic model is adopted, the yield criterion is the Mohr–Coulomb yield criterion, the model Poisson's ratio is $\upsilon = 0.3$, and the initial dynamic shear modulus is $G_{max} = 220$ Mpa. First, the model is analyzed by the initial static analysis. The boundary condition of the model is that the horizontal and vertical directions at the bottom are fixed, the horizontal directions on both sides are fixed, and the vertical directions on both sides are free. Then, the model is calculated and analyzed by equivalent linearity calculation and analysis. The boundary condition of the model is that the horizontal and vertical directions at the bottom are fixed, the vertical directions on both sides are fixed, and the horizontal directions on both sides are free.

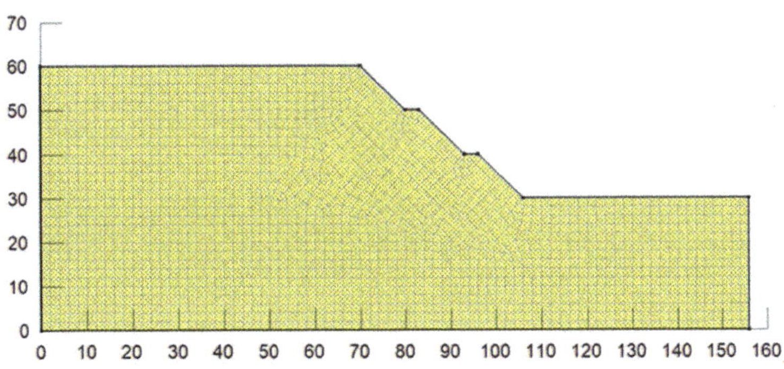

Figure 2. Grid diagram of dynamic analysis and calculation model of multi-stage loess slope under A-1 working condition.

In the dynamic calculation of the model, the EI-Centrol seismic wave with a duration of 30 s and a peak acceleration of 0.2 g is selected for each working condition, which is input at the bottom of the model. The boundary condition of the model is that the horizontal and vertical directions at the bottom of the model are fixed, the vertical directions on both sides of the model are fixed, and the horizontal directions on both sides of the model are free. Because it is defined that the permanent displacement is the horizontal displacement of the slope after experiencing the whole duration range of the earthquake when the safety factor is less than 1.2, the maximum horizontal displacement after the calculation of the model under each working condition is selected, which is the permanent displacement. The calculated permanent displacement of the slope under each working condition in Table 1 is shown in Figure 3.

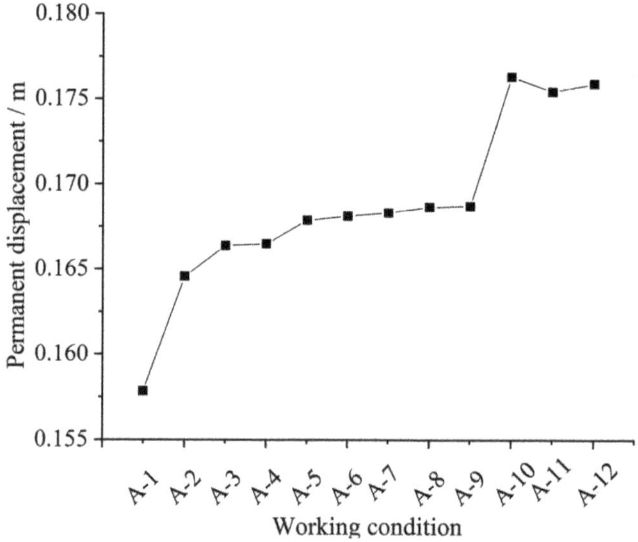

Figure 3. The permanent displacement under various working conditions of Table 1.

As can be seen from Figure 3, when the slope rate of each grade slope changes but the comprehensive slope rate remains unchanged, the minimum horizontal permanent displacement of multi-stage loess slope under the action of earthquake is 0.15784 m under A-1 working condition, and the maximum value is 0.17633 m under A-10 working condition. Additionally, the difference between the maximum permanent displacement and the minimum permanent displacement is 0.01849 m, which is only 10.486% of the maximum permanent displacement. According to the above analysis, it can be shown that it is feasible to use the comprehensive slope rate to calculate the permanent displacement under the condition that the slope rate of each grade changes, but the comprehensive slope rate remains unchanged. It can also be seen from Figure 3 that the maximum horizontal displacement is most directly related to the slope rate of the first-stage slope. When the first-stage slope rate remains unchanged, the difference in the maximum horizontal displacement is very small, such as the working condition of A-10~A-12.

2.2. When the Stage of Multi-Stage Slope Is Changed

In order to explore the influence of the comprehensive slope ratio on the permanent displacement under the condition that the stage of multi-stage loess slope changes and other geometric parameters remain unchanged, the slope model as shown in Table 2 is established. The specific model parameters are the same as those in Section 2.1, only the slope series is changed, and the slope rate of each grade is the same under the same working condition. A total of 20 calculation conditions are set up, the specific conditions are shown in Table 2, and the research results of Zhang [27] are introduced for comparative analysis.

Table 2. Working conditions for calculation of the number of stage changes in multi-stage loess slope.

Working Condition	Slope Stage	Comprehensive Slope Rate	Working Condition	Slope Stage	Comprehensive Slope Rate
B-1	1		B-11	1	
B-2	2		B-12	2	
B-3	3	1:0.75	B-13	3	1:1.25
B-4	4		B-14	4	
B-5	5		B-15	5	
B-6	1		B-16	1	
B-7	2		B-17	2	
B-8	3	1:1	B-18	3	1:1.5
B-9	4		B-19	4	
B-10	5		B-20	5	

The calculated permanent displacement of the slope under each working condition in Table 2 is shown in Figure 4.

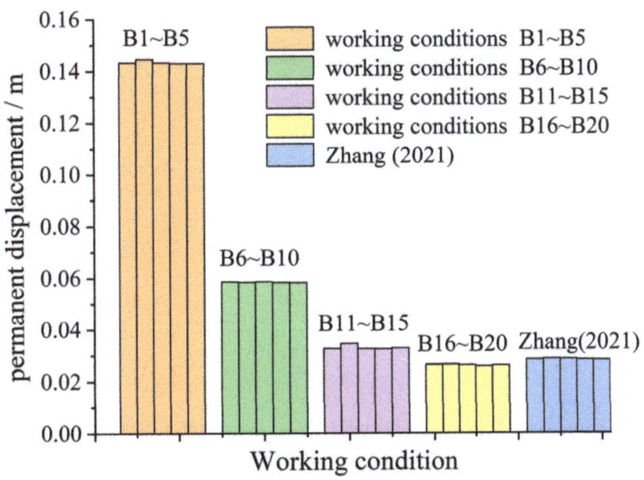

Figure 4. The permanent displacement under various working conditions of Table 2.

It can be seen from Figure 4 that under the condition of constant comprehensive slope rate, the change in the slope series has little influence on the permanent displacement of multi-stage loess slope under the action of earthquake. In the research results of Zhang [27], the total slope angle of the multi-level loess slope model is 41.4°, the width of the platform is 4 m, the total height of the slope is 30 m, and the stage number ranges from 1 to 5. From the research results of Zhang [27], it can also be seen that the change in stage number has little influence on the permanent displacement. Therefore, it is also feasible to use the comprehensive slope rate to calculate the permanent displacement when the stage of multi-stage loess slopes changes, but the comprehensive slope rate remains unchanged.

3. Analysis of Energy Response of Multi-Stage Loess Slope under the Action of Earthquake

Combining the contents of Section 2.1, under the condition that the comprehensive slope rate is constant and other geometric parameters (slope rate of each stage, height of each stage and slope stage of the multi-stage loess slope) change, the variation range of potential sliding soil permanent displacement is small. In order to simplify the calculation of the permanent displacement of the potential sliding soil, the other geometric parameters of the multi-stage loess slope can be replaced by the comprehensive slope rate, and the theoretical analysis and calculation can be carried out. When calculating the permanent displacement of the potential sliding soil, the multi-stage loess slope can also be regarded as a single-stage slope with the same slope rate as the comprehensive slope rate of original slope.

3.1. Basic Assumptions

According to the relevant flow law of rock and soil mass and the principle of limit analysis, in order to simplify the calculation and approach the actual project, the following assumptions are made [23,28,29].

(1) The strength of rock and soil mass obeys the Mohr–Coulomb strength criterion.
(2) When the rock and soil mass fails, the yield surface is convex everywhere and conforms to the relevant flow laws.
(3) When the potential sliding body slides with the lower soil body, its sliding surface is assumed to be a logarithmic spiral along the tangential direction.
(4) When the potential sliding body is in the limit equilibrium state that is about to slide, the acceleration of the potential sliding body is its yield acceleration.
(5) There is no gradual failure of the plastic failure of the soil on the slip surface; that is, the cohesive force of the soil c and the angle of internal friction of the soil φ remain unchanged during the earthquake duration.

3.2. Critical Acceleration of Potential Sliding Soil

As shown in Figure 5, the potential sliding soil ABC appears on the slope surface under the action of earthquake. According to the Newmark slider displacement analysis method, when the soil slides under horizontal and vertical seismic forces, the potential sliding soil ABC will appear in the following four states. The first state is relatively static (the seismic acceleration does not exceed the yield acceleration of the potential sliding soil), the second state is limit equilibrium (the seismic acceleration reaches the yield acceleration of the potential sliding soil), the third state is positive downward sliding (the seismic acceleration exceeds the positive yield acceleration of the potential sliding soil) and the fourth state is reverse upward sliding (the negative seismic acceleration exceeds the negative critical acceleration of the potential sliding soil) [23,28].

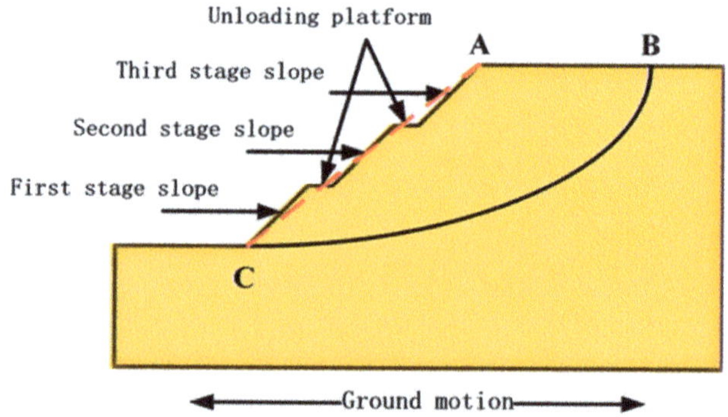

Figure 5. Sliding system of multi-stage loess slope under seismic action.

It is assumed that the comprehensive slope rate of the multi-stage loess slope is 1:n, the inclination between the potential sliding soil and the lower part of the sliding soil is θ, and both horizontal and vertical seismic forces are considered in the theoretical analysis and calculation. When the potential sliding soil ABC is in the limit equilibrium state under earthquake excitation, the positive critical acceleration of the potential sliding soil $a_{cr}(t)$ and the negative critical acceleration $a_{co}(t)$ of the potential sliding soil can be solved by using the upper limit theorem of plastic limit analysis.

3.2.1. Positive Downward Critical Acceleration

When the positive seismic acceleration reaches the critical acceleration of the potential sliding soil, the sliding soil will be in the state of limit equilibrium about to slide at this moment. At this time, the potential sliding soil has a downward velocity v relative to the lower soil, the angle between the velocity and the slip surface is α, and the calculation diagram is shown in Figure 6. According to the theory of plastic limit equilibrium, the external force power $\dot{w}_0$ is equal to the internal force power $\dot{w}_i$ [23].

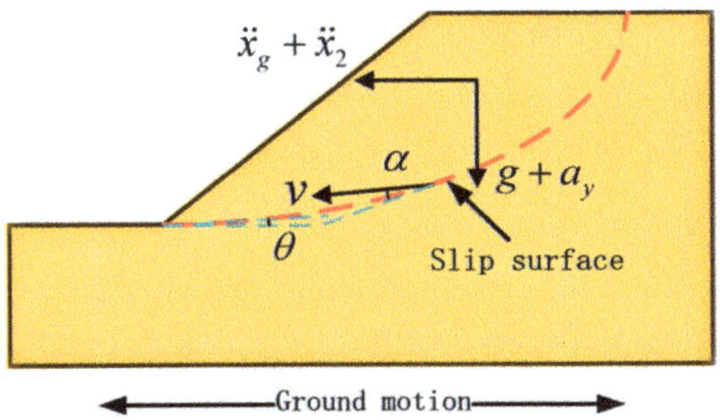

Figure 6. Calculation diagram of positive critical acceleration.

The external force power $\dot{w}_0$ is composed of two parts, namely, the power made by gravity P_w and the power made by seismic inertia force P_i [28]:

$$P_w = m(g + a_y)v\sin(\theta - \alpha) \tag{1}$$

$$P_i = m(\ddot{x}_g + \ddot{x})v\cos(\theta - \alpha) \tag{2}$$

where m is the mass of the potential sliding soil, g is the acceleration of gravity, $\ddot{x}_g$ is the horizontal acceleration of the lower part of the potential sliding soil, and $\ddot{x}$ is the acceleration of the potential sliding soil relative to the lower soil.

The dissipated power of internal force $\dot{w}_i$ is composed of three parts, namely, the energy dissipation of the sliding surface during the motion P_t, the energy dissipation of the potential sliding soil due to the plastic deformation of the soil at the moment of instability P_c, and the energy dissipation caused by the Coulomb damping when the potential sliding soil is in the limit equilibrium state P_d [28]:

$$P_t = clv\cos\varphi \tag{3}$$

$$P_c = csv\cos\varphi \tag{4}$$

$$P_d = mv\left[(g + a_y)\cos\theta - (\ddot{x}_g + \ddot{x})\sin\theta\right]\tan\varphi\cos\alpha \tag{5}$$

According to [23]

$$P_w + P_i = P_t + P_c + P_d \tag{6}$$

The forward critical acceleration of the potential sliding body at this time $a_{cr}(t)$ can be obtained:

$$a_{cr}(t) = \frac{cl\cos\varphi + cs\cos\varphi + m(g+a_y)\cos\theta\tan\varphi\cos\alpha}{m\cos(\theta-\alpha) + m\sin\theta\tan\varphi\cos\alpha} - \frac{m(g+a_y)\sin(\theta-\alpha)}{m\cos(\theta-\alpha)\cos\alpha + m\sin\theta\tan\varphi\cos\alpha} \tag{7}$$

3.2.2. Negative Downward Critical Acceleration

When the seismic acceleration changes from positive to negative, the velocity of the potential sliding soil relative to the lower soil gradually decreases until the seismic acceleration reaches the negative critical acceleration of the potential sliding soil. Then, the potentially sliding soil will be in limit equilibrium state about to slide upward under seismic excitation, and the angle between its velocity v' and the potential sliding surface is β, and the direction is upward. The calculation diagram is shown in Figure 7.

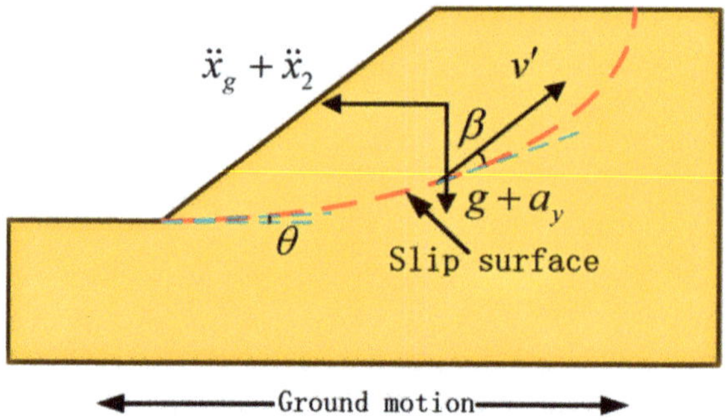

Figure 7. Calculation diagram of negative critical acceleration.

At this time, the power made by the external force $\dot{w}_0$ and the dissipation power of the internal force $\dot{w}_i$ are:

$$\dot{w}_0 = -m(g + a_y)v' \sin(\theta + \beta) + m(\ddot{x}_g + \ddot{x})v' \cos(\theta + \beta) \tag{8}$$

$$\dot{w}_i = clv' \cos\beta + csv' \cos\beta + mv'\left[(g + a_y)\cos\beta + (\ddot{x}_g + \ddot{x})\sin\beta\right]\tan\varphi\cos\beta \tag{9}$$

From $\dot{w}_0 = \dot{w}_i$ [23], the negative critical acceleration $a_{co}(t)$ can be obtained when the potential sliding soil is in the limit equilibrium state, and it should be noted that the negative critical acceleration is a negative value:

$$\begin{aligned} a_{co}(t) = & -\frac{cl\cos\beta + cs\cos\beta + m(g+a_y)\cos\beta\tan\varphi\cos\beta}{m\cos(\theta+\beta)\cos\alpha - m\sin\theta\tan\varphi\cos\beta} \\ & -\frac{m(g+a_y)\sin(\theta+\beta)}{m\cos(\theta+\beta)\cos\alpha - m\sin\theta\tan\varphi\cos\beta} \end{aligned} \tag{10}$$

3.3. *Influence of Inclination Angle of Slip Surface θ and Soil Parameters on Positive and Negative Critical Acceleration*

Since the vertical acceleration of potential sliding soil is no longer a fixed value, it is a more complex value with the change in inclination angle and soil parameters when the inclination angle θ and soil parameters (internal friction angle φ and cohesive force c) are changed. Therefore, the vertical acceleration a_y is not considered in the analysis of this section, which makes it equal to 0. The effects of the inclination angle of the potential slip surface θ and the friction angle of the soil φ and cohesive force c of the soil on the critical acceleration of the potential sliding soil a_c are shown in Figures 8–10. In order to simplify the calculation process, the width of the potential sliding soil is 1 m. The geometric, physical and mechanical parameters of the slope are shown in Table 3.

It can be seen from Figure 8 that as the increase in the inclination angle of the potential slip surface θ, the absolute value of the positive critical acceleration $|a_{cr}(t)|$ decreases with a very small range, which accords with the general law from the point of view of stability. When the inclination angle θ increases from 0 to 40°, the absolute decrease in positive critical acceleration is only 1.07. The absolute value of negative critical acceleration $|a_{co}(t)|$ increases continuously when the inclination angle of potential slip surface θ increases from 0 to 40°, and when the inclination angle of the potential slip surface θ reaches 40°, the absolute value of the negative critical acceleration exceeds 5 g. It can also be seen from Figure 8 that the assumption that the Newmark slider displacement analysis method is not applicable to the sliding soil system with deeper potential slip surface is reasonable. When the slip surface is deep, the negative critical acceleration increases sharply, and it is impossible for the general earthquake and velocity to reach the value shown in Figure 8.

The Newmark slider displacement analysis method does not consider the negative critical acceleration, so it does not consider the reverse displacement.

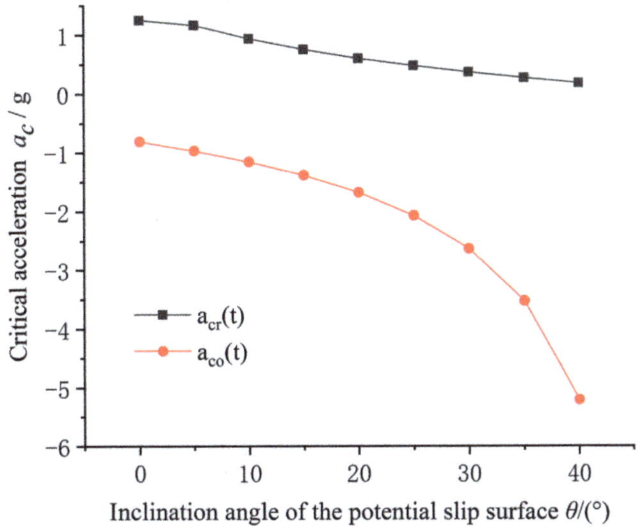

Figure 8. Influence of inclination angle of potential slip surface θ on critical acceleration a_c.

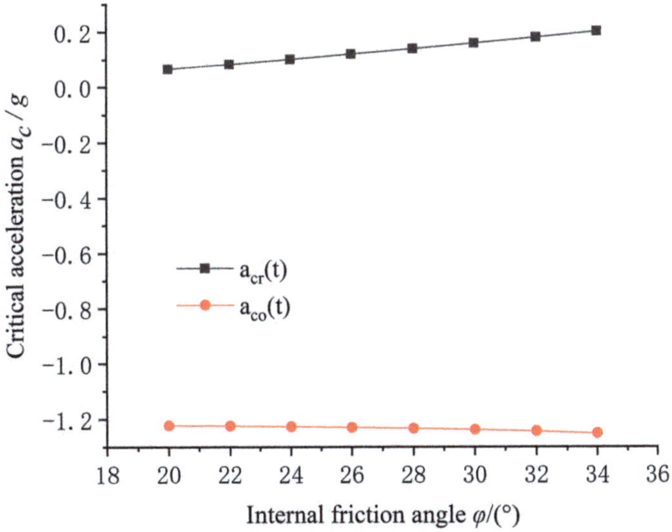

Figure 9. Influence of internal friction angle of soil φ on critical acceleration a_c.

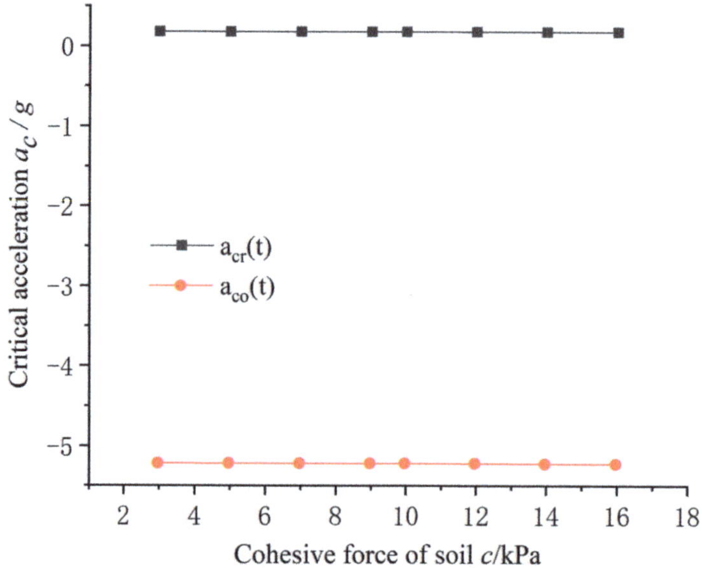

Figure 10. Influence of cohesive force of soil c on critical acceleration a_c.

Table 3. The geometry, physical and mechanical parameters of slope.

Parameters	Length of Potential Slippery Surface l/m	Soil Weight γ/(kN/m^3)	Dynamic Shear Modulus G/Mpa	Cohesive Force c/kPa	Internal Friction Angle φ/(°)	Inclination Angle of Potential Slip Surface θ/(°)
Numerical value	172	16.6	220	15	32	40

It can be seen from Figures 9 and 10 that the friction angle and cohesive force of soil have much less influence on the critical acceleration relative to the inclination angle of the potential sliding surface. It is proved that the energy dissipated by damping and soil deformation is much smaller than the energy dissipated by the motion of the potential sliding soil.

3.4. Influence of Vertical Acceleration on Critical Acceleration in the Action of Earthquake

When an earthquake occurs, both horizontal seismic excitation and vertical seismic excitation act on the slope soil at the same time. Most scholars only consider the influence of horizontal seismic action when considering the dynamic response of the slope under the action of earthquake, but some new studies show that vertical seismic force can change the distribution of dynamic stress in structure, which cannot be ignored in structural seismic and dynamic response analysis [30]. Taking the slope in Table 1 as an example, a monitoring point is set at the top of the slope, and an EI-Cenrol seismic wave with a peak acceleration of 0.2 g is applied at the bottom of the model under the case for considering the vertical acceleration. Then, the horizontal acceleration and the positive critical acceleration within the earthquake duration of 0~5 s is drawn, as shown in Figure 11.

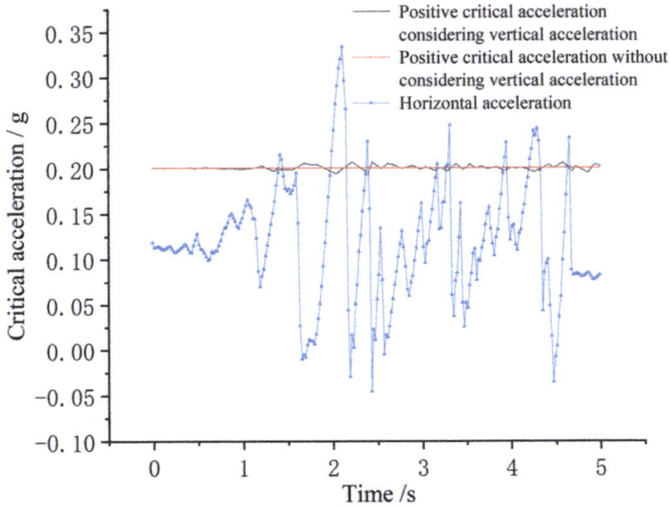

Figure 11. Horizontal acceleration and positive critical acceleration time history.

As shown in Figure 11, in the calculation of the positive critical acceleration, it can be seen from the red line that the positive critical acceleration is a certain value without considering the vertical critical acceleration, which is 0.201 g, and its value is a variable quantity with time when considering vertical acceleration. The magnitude and direction of the vertical acceleration directly affect the magnitude and direction of the critical acceleration. The positive critical acceleration increases when the vertical acceleration is in the same direction as the critical acceleration, and the positive critical acceleration decreases when the vertical acceleration is opposite to the critical acceleration. It can also be seen from Figure 11 that $a_{cr}(t)$ fluctuates up and down around a_{cr} over time, and the number of horizontal acceleration exceeding the critical acceleration is limited. If the vertical acceleration is not considered and the critical acceleration is regarded as a fixed value to calculate the permanent displacement, there may be some errors, which will affect the calculation of the permanent displacement.

3.5. Energy Equation of Sliding Soil System

From the analysis of Figure 6, it can be concluded that under the combined action of horizontal and vertical seismic excitation, sliding soil will move along the potential sliding surface after the acceleration of the sliding soil exceeds the yield acceleration, and the differential equation of motion is:

$$-m(\ddot{x}_g+\ddot{x})\cos\theta - cl - \mu m\left[(g+a_y)\cos\theta + (\ddot{x}_g+\ddot{x})\sin\theta\right] + m(g+a_y)\sin\theta = 0 \quad (11)$$

where m is the friction coefficient between the sliding soil mass and the lower soil mass, $\mu = \tan\varphi$.

Equation (11) can be rearranged:

$$(1+\mu\tan\theta)m(\ddot{x}_g+\ddot{x}) + \frac{cl}{\cos\theta} + \mu m(g+a_y) - m(g+a_y)\tan\theta = 0 \quad (12)$$

During the whole process of the earthquake, the soil of the slope has been in the energy field under the earthquake excitation, and the input, transformation and consumption of energy have been carried out and balanced. On the one hand, the soil is constantly transmitting energy; on the other hand, it continues to consume energy because of the displacement caused by motion. The differential equation of motion of the potential sliding soil is established only in the limit equilibrium state, which can only reflect the energy state of one moment, and cannot represent the process of motion change process of potential sliding soil at other times in the duration range of earthquake. By quantifying the process of energy change in sliding soil in the energy field, the energy response equation of sliding soil under earthquake can be obtained [28,29].

The absolute displacement of potential sliding soil $x_g + x$ at both ends of Equation (12) is integrated in the earthquake duration range [0, *t*], and the absolute energy response equation of potential sliding soil is obtained:

$$\begin{aligned}&\int_0^t (1+\mu\tan\theta)m(\ddot{x}_g+\ddot{x})(\dot{x}_g+\dot{x})dt + \int_0^t \frac{cl}{\cos\theta}(\dot{x}_g+\dot{x})dt + \\ &\int_0^t \mu m(g+a_y)(\dot{x}_g+\dot{x})dt - \int_0^t m(g+a_y)\tan\theta(\dot{x}_g+\dot{x})dt = 0\end{aligned} \tag{13}$$

Simplification can be obtained:

$$\begin{aligned}&\frac{1+\mu\tan\theta}{2}m(\dot{x}_g+\dot{x})^2 + \left[\frac{cl}{\cos\theta} + \mu m_2(g+a_y)\right]x - m(g+a_y)x\tan\theta = \\ &(1+\mu\tan\theta)\cdot\int_0^t m(\ddot{x}_g+\ddot{x})\dot{x}_g dt\end{aligned} \tag{14}$$

where $\dot{x}_g$ is the velocity of the lower soil, x is the displacement of the potential sliding soil relative to the lower soil, that is, the permanent displacement of the potential sliding soil, $\dot{x}_g + \dot{x}$ is the absolute velocity of the potential sliding soil, and $\ddot{x}_g + \ddot{x}$ is the absolute acceleration of the potential sliding soil.

Equation (14) is suitable for any time within the duration range of the earthquake; that is, the values of energy input and output of the whole sliding soil system are always balanced in the earthquake duration range. Figure 12 shows the acceleration of potential sliding soil and lower soil during a certain period of time when the vertical acceleration of the lower soil is not taken into account.

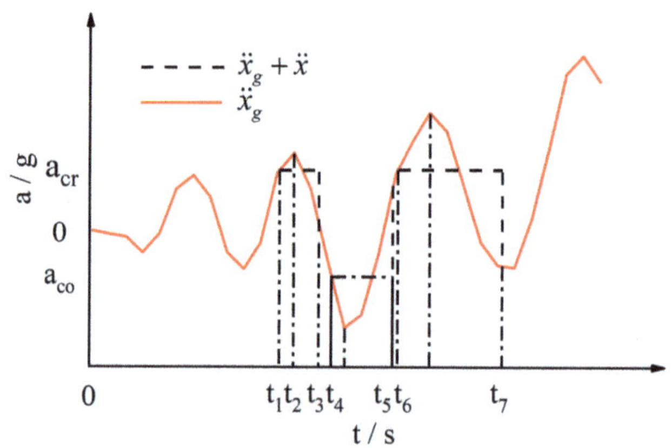

Figure 12. Time history of horizontal acceleration of potential sliding soil and lower soil.

During the time period of [0, t_1], the acceleration of the lower soil mass (the seismic acceleration) does not exceed the positive and negative critical acceleration of the potential sliding soil, and the potential sliding soil remains relatively static with the lower soil. The magnitude of the static friction force at the potential slip surface is equal to the seismic force at the moment, and the direction changes continuously with the direction of the seismic force to counteract the effect of the seismic force. With the change in the direction of the seismic force, the static friction force continues to perform positive and negative work, the kinetic energy of the whole landslide system increases and decreases continuously, and the total energy keeps balance. Additionally, the speed of potential sliding soil is always the same as that of the lower soil. During the time period of [t_1, t_2], the acceleration of the lower soil exceeds the positive critical acceleration a_{cr} of the potential sliding soil, and the potential sliding soil breaks the limit equilibrium state that is about to slide downward and begins to slide downwards. In this process, the acceleration of the lower soil and the sliding soil are $\ddot{x}_g$ and a_{or}, respectively, where the former is accelerated motion, and the latter is uniform acceleration motion. During the time period of $[t_2, t_3]$, the lower soil begins to decelerate, and its acceleration is $\ddot{x}_g$. Due to the action of inertia, the sliding soil still accelerates uniformly, and its acceleration is a_{cr}. At the t_3 moment, the velocity of the lower soil and the sliding soil are consistent again. During the whole time period of $[t_1, t_3]$, The kinetic energy of the system is increased by the work performed by the shear force at the slip surface, the vertical seismic force and the gravity, and the kinetic energy is reduced by the friction work between the lower soil and the sliding soil at the slip surface. At t_4 moment, the acceleration of the lower soil exceeds the reverse critical acceleration a_{or} of the sliding soil, and the sliding soil begins to slide upward. At the t_5 moment, the velocity of the lower soil and the sliding soil are consistent again. Differently from the fact that the work performed by gravity and the shear force at the slip surface increases the system energy in the whole time period of $[t_1, t_3]$, the work performed by gravity and shear force at the slip surface in the whole time period of $[t_4, t_5]$ will dissipate part of the seismic input energy. The movement mode of sliding soil in the time period of $[t_6, t_7]$ is the same as that of $[t_1, t_3]$.

3.6. Critical Input Power and Permanent Displacement

From the term of horizontal seismic input energy at the right end of Equation (14), the instantaneous input energy of the block ΔE_{EQH} with a time interval of Δt within the duration range of the earthquake can be obtained as follows:

$$\Delta E_{EQH} = (1 + \mu \tan\theta) \cdot \int^{\Delta t} m\ddot{x}_z \dot{x}_g dt \tag{15}$$

At this time, the condition of block sliding is that the instantaneous input energy ΔE_{EQH} of the block is greater than the kinetic energy increase in the block in the time period of Δt. After the instantaneous input energy is obtained from Equation (15), the instantaneous input power P_r can be obtained by deriving it:

$$P_r = (1 + \mu \tan\theta) m\ddot{x}_z \dot{x}_g \tag{16}$$

By deriving the first kinetic energy term at the left end of Equation (14), the critical input power of the block P_{cr} can be obtained:

$$P_{cr} = (1 + \mu \tan\theta) m\ddot{x}_z \dot{x}_z \tag{17}$$

When the earthquake input power is less than the critical input power of the block, this part of the input energy can be fully converted and consumed inside the slope, so it will not trigger the block to slide, which is relatively safe. On the contrary, the energy input by ground motion will not be safely converted and consumed. After deriving the kinetic energy term and making it equal to the instantaneous input power of the block P_{res}, the residual input power of the permanent displacement of the block P_{res} can be obtained:

$$P_{res} = (1 + \mu \tan\theta) m\ddot{x}_z (\dot{x}_g - \dot{x}_z) \tag{18}$$

The residual input power P_{res} is integrated within $[0, t]$ during the duration of the earthquake, the part of the residual power input energy of the block under earthquake action can be obtained, which is defined as the residual energy of the block E_{RES} [29]:

$$E_{RES} = (1 + \mu \tan\theta) m \int_0^t \ddot{x}_z (\dot{x}_g - \dot{x}_z) dt \tag{19}$$

From Equations (14) and (19), the residual energy E_{RES} can be defined as the difference between the absolute input energy of horizontal earthquake E_{EQH} and the absolute kinetic energy of sliding soil E_K, that is, $E_{RES} = E_{EQH} - E_K$. The generation of residual energy is the trigger condition of the potential sliding of sliding soil, and contributes to the generation of permanent displacement of sliding soil together with the increase or decrease in potential energy produced by sliding of soil and the input energy of the vertical seismic wave. Considering the dynamic friction coefficient of sliding soil $\mu = \tan\varphi$, according to the definition of residual energy and Equation (19), the following results can be obtained:

$$\left[\frac{cl}{\cos\theta} + m(g + a_y)(\tan\varphi - \tan\theta)\right]x = (1 + \tan\varphi\tan\theta)m\int_0^t \ddot{x}_z(\dot{x}_g - \dot{x}_z)dt \quad (20)$$

According to Equation (20), the permanent displacement x of the block in the whole earthquake duration range can be obtained.

4. Example Verification

4.1. Model Parameters

It is known that the total height of three-stage homogeneous loess slope is 30 m, the height of each slope is $h_1 = h_2 = h_3 = 10$ m, the slope rate of each grade is $i_1 = i_2 = i_3 = 1{:}0.75$, the width of the platform is $B_1 = B_2 = B_3 = 2$ m, the friction angle of soil is $\varphi = 32°$, the cohesive force is $c = 15$ kPa, and the natural heavy of soil is $\gamma = 16.6$ kN/m^3. The safety grade of the slope is grade II, and the seismic fortification intensity in this area is VIII.

4.2. Comparison of Calculation Results

The calculated time history of positive critical acceleration and horizontal acceleration is shown in Figure 13. The numerical simulation adopts EI-Centrol seismic wave with horizontal peak acceleration of 0.2 g and vertical peak acceleration of 0.2 g, and the duration is 30 s. The positive and negative critical acceleration time histories of sliding soil are calculated by Equations (7) and (10), respectively, and the average values of positive and negative critical acceleration are 0.185 g and −5.214 g, respectively. The permanent displacement calculated in this paper is compared with that calculated by PLAXIS 3D software and Geo-Studio software, and the comparison results are shown in Table 4.

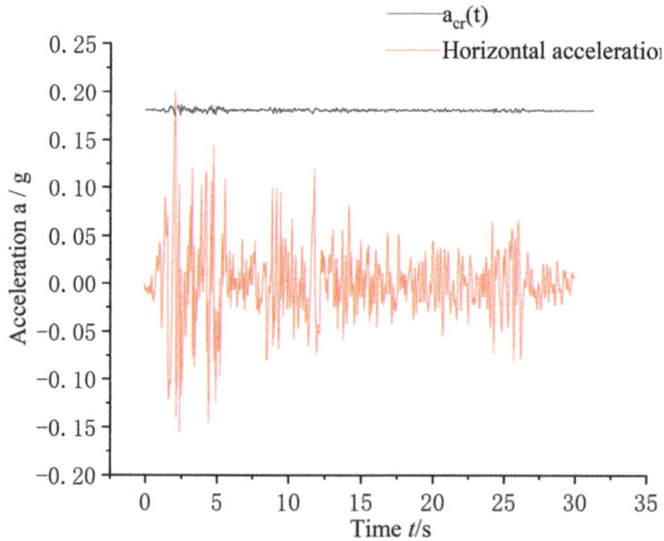

Figure 13. The horizontal acceleration and positive critical acceleration of the calculation model.

Table 4. Comparison of permanent displacement of sliding soil.

Calculation Method	Permanent Displacement/m
The method of this paper	0.186
The method of PLAXIS 3D	0.118
The method of GEO-Studio	0.23

It can be seen from Figure 13 that the calculated value of negative critical acceleration is much larger than the peak value of horizontal acceleration, so it can be concluded that the negative critical acceleration has no effect on permanent displacement. As can be seen from Table 4, the permanent displacement calculated by this paper is larger than that calculated by PLAXIS 3D and smaller than that calculated by GEO-Studio, which is closer to that calculated by GEO-Studio. This is because the Geo-Studio software is a two-dimensional calculation software and is relatively simplified in the calculation process, and the derivation process of this paper does not consider the influence of the lateral deformation of soil on the calculation of permanent displacement. To a certain extent, the calculation method in this paper is reliable.

4.3. Influence of the Change in Gravity Potential Energy on Permanent Displacement

In order to explore the influence of the reduction in gravity potential energy on the permanent displacement of the slope, the inclination angle of the slip surface is changed without considering the vertical acceleration, the critical acceleration, permanent displacement, gravity potential energy drop and seismic input energy of sliding soil are calculated by using the derived theoretical formulas, as shown in Table 5. The ratio of the decrease in seismic input energy to gravitational potential energy and the variation of permanent displacement with the inclination angle of slip surface are calculated, as shown in Figure 14.

Table 5. Critical acceleration, permanent displacement and gravitational potential energy drop of sliding soil.

The Inclination Angle of the Slip Surface θ/(°)	Critical Acceleration		Permanent Displacement x/m	Gravitational Potential Energy/kJ	Seismic Input Energy/kJ
	Positive Critical Acceleration a_{cr}/g	Negative Critical Acceleration a_{co}/g			
0	1.25	−0.81	0.003	0	4.8
5	1.16	−0.97	0.006	15.8	12.9
10	0.93	−1.16	0.029	132	35.7
15	0.75	−1.39	0.058	458	86.4
20	0.60	−1.68	0.147	1247	143.2
25	0.48	−2.07	0.238	2956	175.8
30	0.37	−2.63	0.445	7234	239.6
35	0.27	−3.53	1.175	26472	257.7
40	0.18	−5.21	1.864	125479	277.4

It can be seen from Table 5 and Figure 14 that with the increase in the inclination angle of slip surface, the permanent displacement changes greatly at 30°, and it suddenly increases sharply from the steady increase before 30°. Meanwhile, the ratio of seismic input energy to gravitational potential energy drop decreases sharply after 10°, and the absolute values of positive critical acceleration and negative critical acceleration decrease and increase, respectively. When the inclination angle of the potential sliding surface is greater than 15°, the contribution of the seismic input energy to the permanent displacement of the slope under the action of the earthquake is relatively limited. It shows that for the permanent displacement of the slope excited by earthquake, earthquake is only a triggering condition and factor, and the real factor causing permanent displacement is gravitational potential energy.

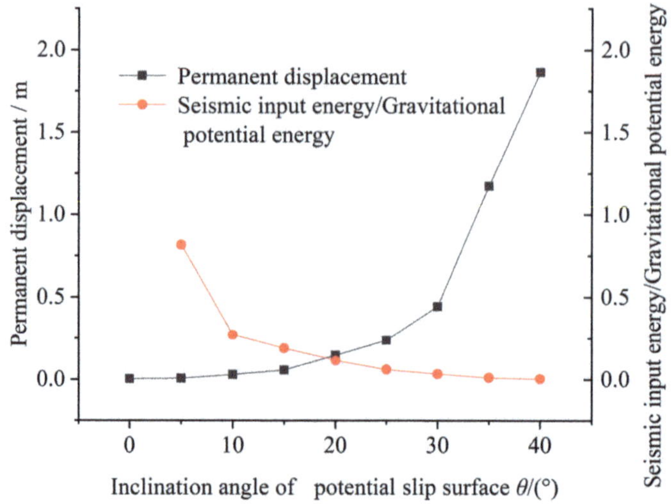

Figure 14. The variation curve on the ratio of seismic input energy to gravitational potential energy and the permanent displacement with inclination angle of potential slip surface.

5. Conclusions

1. Through a large number of numerical simulation tests, it is concluded that the geometric characteristics of the slope cannot be taken into account to a certain extent when calculating the permanent displacement of sliding soil of multi-stage loess slope under the action of earthquake; in addition, it is feasible to use the comprehensive slope to calculate the permanent displacement of the sliding body of multi-stage loess slope under the action of earthquake.
2. On the premise of using the comprehensive slope rate to calculate the permanent displacement of multi-stage loess slope under earthquake, based on the energy method, combined with the Newmark slide displacement method, considering soil deformation and damping, the formulas for calculating positive and negative critical acceleration and permanent displacement of sliding soil are reasonable.
3. The critical acceleration of the sliding soil is mainly affected by the inclination angle of the sliding surface, but not by the soil parameters. The earthquake is only an inducing factor for the permanent displacement of the sliding soil, and the decrease in gravitational potential energy plays a key role in the permanent displacement.
4. The method derived in this paper comprehensively considers the influence of negative critical acceleration on the solution of permanent displacement when the inclination angle of slip surface is small and the seismic acceleration is large, and based on the principle of energy balance in the process of calculation and derivation, the calculation results are thus more accurate.

Author Contributions: Conceptualization: S.Y.; data curation: X.Z. (Xiaobing Zhang); formal analysis: X.Z. (Xiaobing Zhang), X.Z. (Xin Zhang) and S.Y.; investigation: X.Z. (Xiaobing Zhang), X.Z. (Xin Zhang) and S.Y.; methodology: X.Z. (Xiaobing Zhang) and S.Y.; software: X.Z. (Xiaobing Zhang); writing—original draft: X.Z. (Xiaobing Zhang) and S.Y.; writing—review and editing: X.Z. (Xiaobing Zhang), X.Z. (Xin Zhang) and S.Y. All authors have read and agreed to the published version of the manuscript.

Funding: This research was funded by the National Natural Science Foundation of China (No. 52168050) and the National Natural Science Foundation of China (No. 51768040).

Institutional Review Board Statement: Not applicable.

Informed Consent Statement: The authors listed in this paper all agree to publish this paper.

Data Availability Statement: The data of simulation and theoretical calculation in the article are not freely available due to legal concerns and commercial confidentiality. Nevertheless, all the concepts and procedures are explained in the presented research, and parts of the research may be available upon request.

Acknowledgments: The authors of the present work feel grateful and would like to thank Northwest Research Institute Co., Ltd. of C.R.E.C, Shandong Jianzhu University, National Natural Science Foundation of China, Lanzhou University of Technology, for providing financial and technical support in research.

Conflicts of Interest: The authors declare that they have no conflict of interest.

References

1. Xie, Y.L.; Liu, X.R.; Yan, C.G.; Yang, Z.P.; Li, J.C.; Zhou, Z.J.; Yue, X.B. Research progress of special soil and rock engineering slopes. *China Civ. Eng. J.* **2020**, *53*, 93–105.
2. Li, Y.R.; Shi, W.H.; Aydin, A.; Beroya-Eitner, M.A.; Gao, G.H. Loess genesis and worldwide distribution. *Earth-Sci. Rev.* **2020**, *201*, 1–22. [CrossRef]
3. Bai, B.; Nie, Q.; Zhang, Y.K.; Wang, X.L.; Hu, W. Cotransport of heavy metals and SiO_2 particles at different temperatures by seepage. *J. Hydrol.* **2021**, *597*, 125771. [CrossRef]
4. Ye, S.H.; Huang, A.P. Sensitivity of high fill slope stability factors under seismic conditions. *Soil Mech. Found. Eng.* **2020**, *57*, 356–363.
5. Xu, L.; Gao, C.; Lan, T.; Lei, J.; Zuo, L. Influence of grading on the compressibility of saturated loess soils. *Geotech. Lett.* **2020**, *10*, 198–204. [CrossRef]
6. Tian, W.T.; Sun, J.J.; Wang, L.M.; Xu, S.H.; Liu, K.; Sun, Y. Research progress and frontier scientific problems in loess dynamics. *Chin. J. Geotech. Eng.* **2015**, *37*, 2119–2127.
7. Yan, Z.X.; Guo, B.; He, X.; Jiang, P. Study of effect of platform width on dynamic response and failure mechanism of stepped slopes under earthquake. *Rock Soil Mech.* **2012**, *33*, 352–358.
8. Zhang, F.; Leshchinsky, D.; Gao, Y.F.; Yang, S.C. Corner reinforced slopes: Required strength and length of reinforcement based on internal stability. *Geotext. Geomembr.* **2019**, *47*, 408–416. [CrossRef]
9. Ji, J.; Zhang, C.S.; Gao, Y.F.; Kodikara, J. Effect of 2D spatial variability on slope reliability: A simplified FORM analysis. *Geosci. Front.* **2018**, *9*, 1631–1638. [CrossRef]
10. Bai, B.; Wang, Y.; Rao, D.Y.; Bai, F. The effective thermal conductivity of unsaturated porous media deduced by pore-scale SPH simulation. *Front. Earth Sci.-Witz.* **2022**, *10*, 943853. [CrossRef]
11. Newmark, N.M. Effects of Earthquakes on Dams and Embankments. *Géotechnique* **1965**, *15*, 139–160. [CrossRef]
12. Makdisi, F.I.; Seed, H.B. Simplified procedure for estimation dam and embankment earthquake induced deformations. *Geotech. Eng. Div.* **1978**, *104*, 849–868. [CrossRef]
13. Steven, L.K. Modified Newmark Model for Seismic Displacements of Compliant Slopes. *J. Geotech. Geoenviron.* **1997**, *123*, 635–644.
14. Ellen, M.R.; Jonathan, D.B. An examination of simplified earthquake-induced displacement procedures for earth structures. *Can. Geotech. J.* **1999**, *36*, 72–87.
15. Ellen, M.R.; Jonathan, D.B. Nonlinear coupled seismic sliding analysis of earth structures. *Can. Geotech. J.* **2000**, *126*, 1002–1014.
16. Bai, B.; Zhou, R.; Cai, G.; Hu, W.; Yang, G. Coupled thermo-hydro-mechanical mechanism in view of the soil particle rearrangement of granular thermodynamics. *Comput. Geotech.* **2021**, *137*, 104272. [CrossRef]
17. Zhang, N.; Zhang, Y.; Gao, Y.; Pak, R. An exact solution for SH-Wave scattering by a radially multi-layered inhomogeneous semi-cylindrical canyon. *Geophys. J. Int.* **2019**, *217*, 1232–1260. [CrossRef]
18. Zhang, N.; Zhang, Y.; Gao, Y.F.; Pak, R.Y.S.; Yang, J. Site amplification effects of a radially multi-layered semi-cylindrical canyon on seismic response of an earth and rockfill dam. *Soil Dyn. Earthq. Eng.* **2019**, *116*, 145–163. [CrossRef]
19. Bray, J.D.; Travasarou, T. Simplified procedure for estimating earthquake-induced deviatoric slope displacements. *J. Geotech. Geoenviron. Eng.* **2007**, *133*, 381–392. [CrossRef]
20. Hsieh, S.Y.; Lee, C.T. Empirical estimation of the Newmark displacement from the Arias intensity and critical acceleration. *Eng. Geol.* **2011**, *122*, 34–42. [CrossRef]
21. Du, W.Q.; Wang, G.; Huang, D.R. Influence of slope property variabilities on seismic sliding displacement analysis. *Eng. Geol.* **2018**, *242*, 121–129. [CrossRef]
22. Song, J.; Fan, Q.Q.; Feng, T.G.; Chen, Z.Q.; Chen, J.; Gao, Y.F. A multi-block sliding approach to calculate the permanent seismic displacement of slopes. *Eng. Geol.* **2019**, *255*, 48–58. [CrossRef]
23. Ye, S.H.; Zhao, Z.F. Allowable displacement of slope supported by frame structure with anchors under earthquake. *Int. J. Geomech.* **2020**, *20*, 04020188. [CrossRef]
24. Ye, S.H.; Fang, G.W.; Zhu, Y.P. Model Establishment and Response Analysis of Slope Reinforced by Frame with Prestressed Anchors under Seismic Considering the Prestress. *Soil Dyn. Earthq. Eng.* **2019**, *122*, 228–234. [CrossRef]

25. Ye, S.H.; Zhao, Z.F. Seismic response of pre-stressed anchors with frame structure. *Math. Probl. Eng.* **2020**, *2020*, 1–15.
26. Ye, S.H.; Zhao, Z.F.; Zhu, Y.P. Dynamic Response Analysis Of Loess Slope Reinforced By Frame Anchors Based on Numerical Simulation And Shaking Table Test. *J. Geoengin.* **2020**, *15*, 89–101.
27. Zhang, R.H. Stability Analysis of Multistage Loess Slope under Earthquake Action. Master's Thesis, Lanzhou University of Technology, Lanzhou, China, 2021.
28. Kokusho, T. Energy-Based Newmark Method for earthquake-induced slope displacements. *Soil Dyn. Earthq. Eng.* **2019**, *121*, 121–134. [CrossRef]
29. Korzec, A.; Jankowski, R. Extended Newmark method to assess stability of slope under bidirectional seismic loading. *Soil Dyn. Earthq. Eng.* **2021**, *143*, 106600. [CrossRef]
30. Shukha, R.; Baker, R. Design implications of the vertical pseudo-static coefficient in slope analysis. *Comput. Geotech.* **2008**, *35*, 86–96. [CrossRef]

applied sciences

MDPI

Article

Study on Share Rate of Support Structure for Super-Large Span Twin Tunnels with Small Interval

Xuwei Zhao [1], Keguo Sun [1,*], Yingzhou Zhen [1], Yiqin Hong [2] and Huichao Zhou [3]

1 Key Laboratory of Transportation Tunnel Engineering, Southwest Jiaotong University, Chengdu 610031, China; zxw@sty.sh.cn (X.Z.); zhenyz2022@126.com (Y.Z.)
2 China Southwest Architectural Design and Research Institute Corp., Ltd., Chengdu 610095, China; hongyiqin2022@126.com
3 The 3rd Railway Survey and Design Institute Group Corporation, Tianjin 300251, China; zhouhc2022@126.com
* Correspondence: sunkeg@126.com

Abstract: Super-large span twin tunnels, which can relieve traffic burden and improve the efficiency of roads, have been rapidly popularized in urban construction. Hence, it is very important to master the share rate of support structure in this sort of tunnel correctly. In this paper, some field monitoring and numerical simulation work has been carried out in such a tunnel in Daling, China. The content of field monitoring consisted of the vault settlement, stress of the concrete, and the sum pressure. Firstly, the calculation formula of rock pressure in this kind of tunnel had been deduced, which was based on the theoretical sliding failure mode. Then, the calculated rock pressure was applied to the load structure model step by step with 10% intervals (10%, 20%, ...). By comparing the monitoring and the simulation results on vault settlement, the share rate of the primary support was ensured to be 40%. Similarly, by comparing the result on safety factors, the share rate of the secondary lining was obtained (40%). The remained load was born mainly by the surrounding rock itself and the advanced support. Finally, the reliability of the results were verified by the sum pressure. This paper can provide reference for the effective design of support structure in super-large span twin tunnels.

Keywords: twin tunnels; super-large span; share rate; field monitoring; finite element method

check for updates

Citation: Zhao, X.; Sun, K.; Zhen, Y.; Hong, Y.; Zhou, H. Study on Share Rate of Support Structure for Super-Large Span Twin Tunnels with Small Interval. *Appl. Sci.* **2022**, *12*, 7498. https://doi.org/10.3390/app12157498

Academic Editor: Bing Bai

Received: 29 June 2022
Accepted: 23 July 2022
Published: 26 July 2022

1. Introduction

Recently, with the proceeding of urbanization, the demand from traffic is surging. As a result, the need for improving the capacity of the transportation system, specifically the cross section of tunnels in cities, is getting larger to contain four or more lanes [1]. The tunnels with large-span cross section, however, are usually oblate with a small rise-span ratio, which is challenging to design them in a safe and reliable way. At the same time, on the limitation of land resources and route optimization, twin tunnels with small intervals are becoming more usual at the portals, whose mechanical behaviour is much more complicated than ordinary tunnels.

There are many studies on large-span tunnels and ordinary twin tunnels, respectively. In terms of large-span tunnels, Feng Jimeng [2] evaluated the different installation timing of initial ground support for large-span tunnel in hard rock. Hou Fujin [3] and others [4–7] studied the excavation (drill and blast) and construction of large-span tunnels and mechanical characteristics of primary support. Li Shucai [8] used hydro-mechanics to research the influence of groundwater on the surrounding rock of large-span tunnels.

As for the ordinary, Chortis, F. [9] quantifies the interaction effect and preliminarily estimates the main support requirements of twin tunnels based on the stress state of a single tunnel, and provided nomographs for preliminary assessments of the optimum spacing between twin tunnels. Vinod and others [10,11] also studied the influence of shape and interval on twin tunnels and sandwiched rock. Boon Chiaweng [12] researched the mechanics performance of a twin arch tunnel under different rock covers. Golshani and others [13–15] found a method for selection of optimum distance between tunnel tunnels under different conditions. He Chao and others [15] studied the vibration of twin tunnels and found that certain factors (such as the soil and the interval) can affect it. Khabbaz and others [16–20] researched and monitored the twin tunnels under different conditions. Bai Bing and others [21–28] have studied the interaction between the law of rock and soil and structural characteristics of constructions. Besides, Li Pengfei and others [29–31] studied the surrounding rock pressure of deep twin tunnels. And Zhong Zuliang and others [32–34] researched the rock pressure of shallow-buried twin tunnels.

Nonetheless, for large-span twin tunnels there are few studies regarding it. Cui S. [35] analysed the mechanical mechanism of highway tunnels with large span and small spacing during the construction and found that large-span twin tunnels were hard to construct, depending on the excavation area, the thickness, and the attributes of the sandwiched rock and the terrain.

The engineering of super large-span twin tunnels, which can contain eight lanes or more, is not usual at home and abroad. Although the total load of surrounding rock is almost explicit nowadays, its distribution law of surrounding rock pressure as well as the share rate of every part of the retaining structure has not yet been considered clearly. However, designs of primary support and secondary lining are respective, so finding out each part's specific share rate of load is logical for design and construction of actual engineering project. Therefore, in this paper, based on the theory of stratigraphic sliding failure mode and the outcome of monitoring and measurement in the Daling tunnel (twin tunnels, four lanes in each hole), the distribution law of surrounding rock pressure and the share rates of advance support, primary support, and secondary lining are studied. Eventually, by comparing and analyzing the theory and practical results, the correctness of the research method is ensured, which can provide reference for the design and optimization of the retaining structure of super large-span twin tunnels, especially 8-lane (or more) twin tunnels in cities.

2. Methodology

The aim of this research is to find the share rate of the support structure for twin tunnels with super-large span, especially the primary support and the secondary lining. Therefore, the outcomes of numerical simulation and field measurement are compared and the conclusion is verified. In this paper, Daling Tunnel in Jinan City was simulated and monitored for achieving the goals. The whole process is shown in Figure 1.

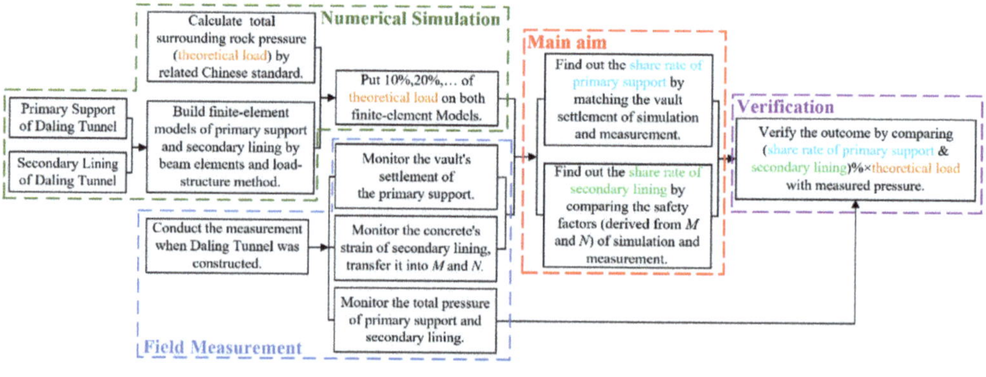

Figure 1. The process of this research. Numerical simulation and field measurement are combined to find out the share rates of primary support and secondary lining. Verification is proposed in final phase.

2.1. Daling Tunnel

Daling Tunnel, located in the southeast of Jinan City, is a twin tunnel with a super-large span cross section, which was monitored in this research. This twin tunnel contains 8 lanes (4 lanes each) with a design speed of 80 km/h, which has great capacity in transportation. The height of the twin tunnel is 12.9 m and the span is 19.4 m. Additionally, the net distance between the twin tunnel is 12.6 m.

Daling Tunnel was excavated by the drill-blast method and supported by the composite lining, which consisted of primary support and secondary support. Furthermore, the whole construction process of this tunnel adhered to the principles of NATM. In terms of the composite lining, the primary support included C25 shotcrete, welded-wire fabric reinforcement, steel arches, and hollow rock bolts (used for grouting and reinforcing). The secondary lining is made mainly of the cast-in-place reinforced concrete (C30). Furthermore, the advance support mainly consists of supporting sheds and small grouting pipes. The information of Daling Tunnel is shown in Figure 2.

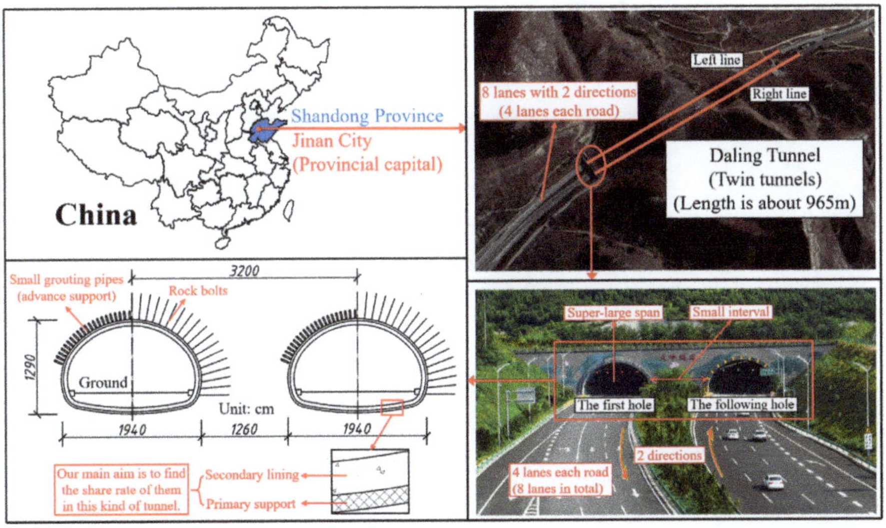

Figure 2. Information of Daling Tunnel. For optimal traffic capacity, these twin tunnels have super-large span and small interval. The main body of them are both composite lining and bolts.

The cross section of Daling Tunnel we measured and monitored is Level-V surrounding rock [1], which is partly soft and cracked. The related physical parameters are shown in Table 1.

Table 1. The related parameters of Daling Tunnel.

Parameter	Value
Unit weight, γ (kN/m^3)	20
Simplified internal friction angle, φ_c	45°
Real internal friction angle, φ	22°
Friction angle of the sliding plane, θ	13.2° ($0.6 \times \varphi$)
Span, B (m)	19.4
Height, H_l (m)	12.9
Net interval, D (m)	12.6
Buried depth, H (m)	20

2.2. Theoretical Load

Firstly, we need to assume that the load's distribution of the super-large span twin tunnels is similar to the normal tunnel, which can be calculated referring to the standard [1]. Then, via the initial formulas of related Chinese standard and various deductions, the total surrounding rock pressure (theoretical load) of the twin tunnels was assured. The details of the calculation and deduction are shown in Section 3.

Next, all the parameters were put into the formulas in Section 3, and the results are shown in Figure 3. In Figure 3, the surrounding rock pressure of the first hole (the hole excavated firstly) and the following hole (the hole excavated later) was symmetric. For this reason, in the following, the analysis will focus on the first hole, while the following hole will be skipped.

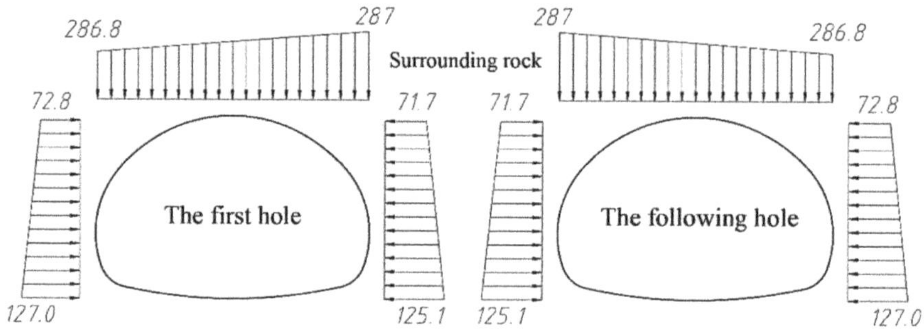

Figure 3. Theoretical load of Daling Tunnel (unit: kPa). The process of calculation will be mentioned in Section 3.

2.3. Finite-Element Model

In this paper, to find out the share rates of support structure, numerical simulation was used for matching the results of field measurement. ANSYS, a powerful finite-element software, was used to build simulation models of primary-support and secondary-lining models of Daling Tunnel, which were both built by beam elements (Beam 3). This is a kind of 2-D element, which has 3-degrees of freedom on each node and can endure axial tension or compression. The parameters of primary support and secondary lining are shown in Table 2.

Table 2. The parameters of primary support and secondary lining.

Part	Young's Model /GPa	Poisson's Ratio	Unit Weight /kN/m^3	Thickness /m
Primary Support	25.3	0.2	25	0.3
Secondary Lining	28	0.2	25	0.7

As for the boundary conditions, the springs were built by link elements (Link 10) and used to simulate the effect of the surrounding rock. Link 10 is a kind of 3-D two-node element, which has 3-degrees of freedom on each node. Considering the mechanics characteristics of the surrounding rock, these springs were set to only be compressed. Furthermore, the bases of springs were completely fixed. According to the parameters of surrounding rocks in Table 1, the length and elastic module of springs are 0.2 m and 200 MPa, respectively.

In terms of load, the form and direction of load is the same as the theoretical load (shown in Figure 3). Furthermore, the value of load is a portion (10%, 20%, ...) of the theoretical load. Moreover, the gravity is 9.8 m/s^2 downward and applied to all elements. The model is shown in Figure 4; (Since the models of the primary support and secondary lining are similar, only one model is shown as an example.)

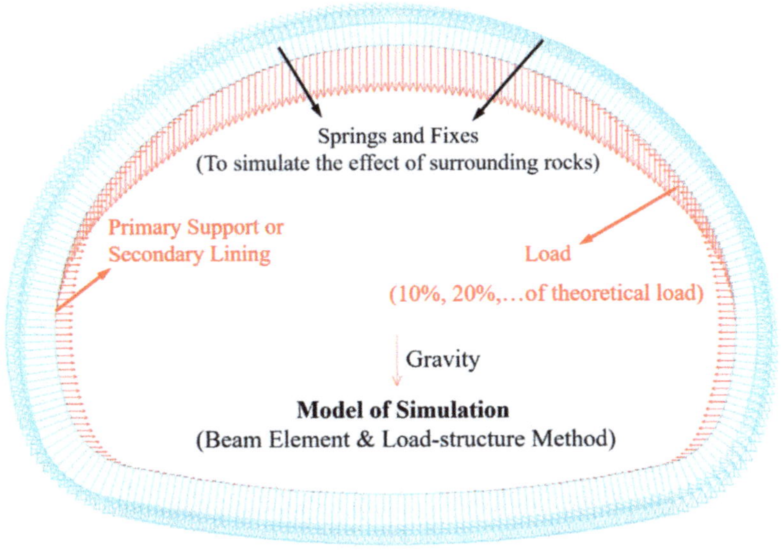

Figure 4. The finite-element model of simulation. It is a load-structure model and mainly simulated by beam elements and springs.

2.4. Field Measurement

In this research, three items were monitored in the first hole of Daling Tunnel. The arrangement of measure points is shown in Figure 5. In Figure 5, the rock bolts are omitted.

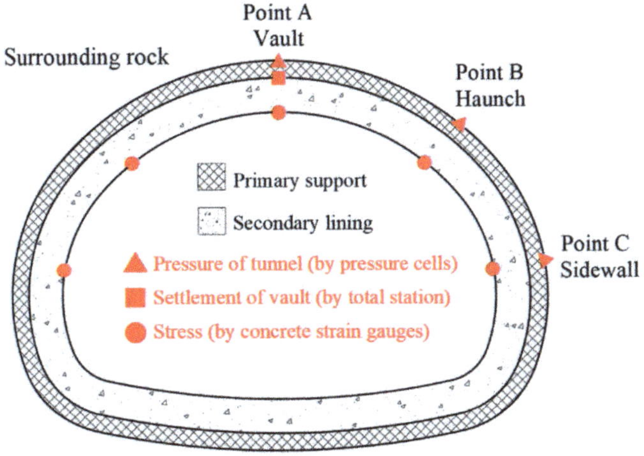

Figure 5. The measure points of Daling Tunnel. Three items were observed in the field measurement.

(1) Settlement of the vault of primary support was measured by total stations and steel rulers to find out the share rate of primary support.
(2) Stress of the secondary lining (derived from the strain) was measured by concrete strain gauges. The gauges were put on both sides of the secondary lining at each point. This is for finding out the share rate of secondary lining. The arrangement of concrete strains is shown in Figure 6.

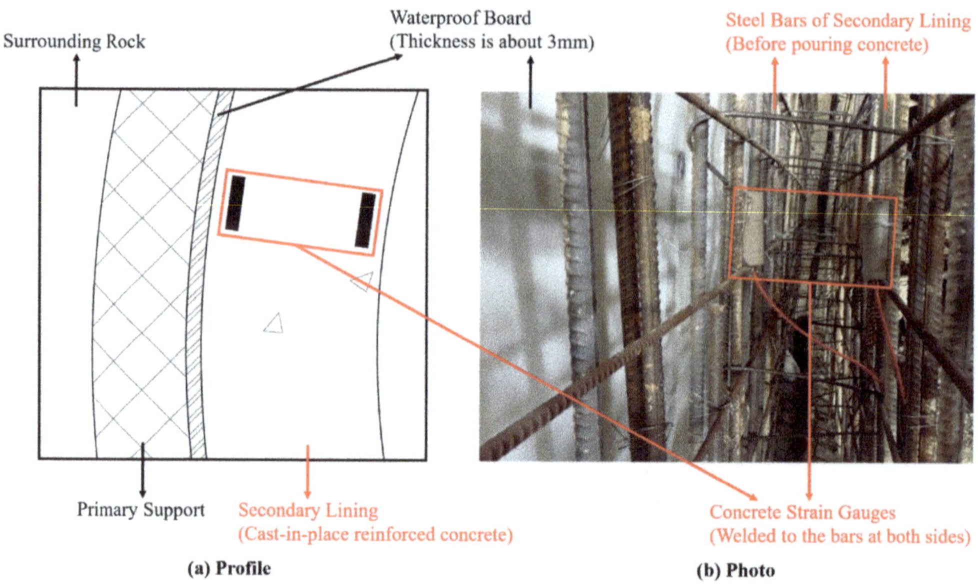

Figure 6. Concrete strain gauges in secondary lining. The most internal force of secondary lining can be obtained by the gauges.

(3) Total pressure of the primary support and secondary lining was measured by pressure cells arranged at the points A, B, and C at the outside of the primary support. This is for comparison and verification. The arrangement of pressure cells is shown in Figure 7.

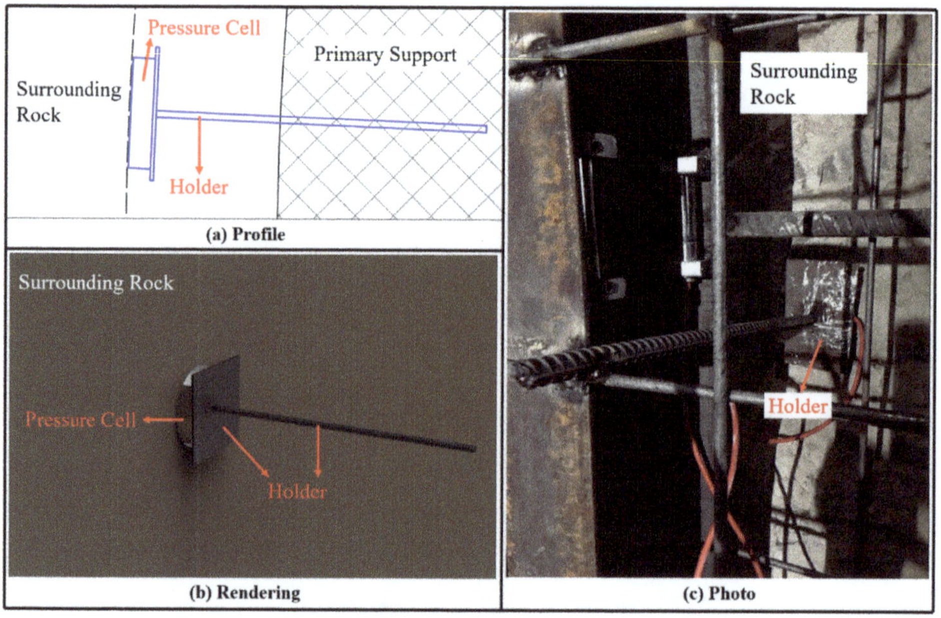

Figure 7. Pressure cells between surrounding rock and primary support. By these cells, the total pressure of the primary support and secondary lining can be measured.

3. Theoretical Calculation of Surrounding Rock Pressure

3.1. Failure Mode and Assumptions

In Chinese standard [1], the method of calculating surrounding rock pressure of normal tunnels (usually including two or three lanes) is given. This method can be generalized to super-large span twin tunnels (four lanes each hole, wider than a normal tunnel). The failure mode, necessary assumptions, and theoretical method are introduced below.

The theoretical sliding failure mode of the surrounding rock of the twin tunnels is shown in Figure 8. W_1, W_2, and W_3 are the gravity of the rock mass $A'C'E'$ (ACE), $IKOM$, and $I'KOM'$, respectively. T_1 and T_2 are the frictional force caused by the soil on both sides of the tunnel to resist the settlement of the vault. Furthermore, φ_c is the simplified internal friction angle of the surrounding rock. According to the Chinese standard, φ_c is usually greater than the real internal friction angle (φ). When using φ_c, it is valid regardless of the cohesion force). θ is determined by referring to the Chinese standard. Here are some explanation and hypotheses for this sliding failure mode:

(1) Suppose that the ground was horizontal, the rock mass was homogenous, and isotropic and the twin tunnel was symmetry and parallel. Moreover, excavation of left and right holes is sequent and full section.

(2) The excavation of the first hole is similar to an ordinary single-hole tunnel. It means that the fracture planes on both sides of the first hole, which are shown as $A'C'$ and $M'J'$ in Figure 8, are two inclined straight planes, and at an angle of β_1 to the horizon. In addition, the pressure inside and outside (shown in Figure 8) is symmetric.

(3) When the following hole is excavated, the fracture plane at the outside of the hole (AC) is at angle β_1 to the horizon, while at the inside of the hole the angle of the fracture plane (MO) is assumed as β_2. Focus on the triangle OJJ', when the following hole is excavated, it is inclined to slide down along the plane JM. However, since the excavation of the first hole has induced a relative slippage at the plane OJ' and undermines the cohesion force along the plane OJ', usually the triangle OJJ' will not slide and fracture along the plane JO. Instead, the tensile fracture plane will be formed in the triangle OJJ', which is assumed as a vertical plane (OK). In summary, the fracture plane at the inside of the following hole is assumed as KOM.

(4) According to the sliding trend of the triangle OJJ' caused by the sequent excavation of the twin tunnel, and based on the theory of soil mechanics, the interactive force (N) in normal direction at the fracture plane OK must be less than the earth pressure at-rest. For safety, let N equal 0.

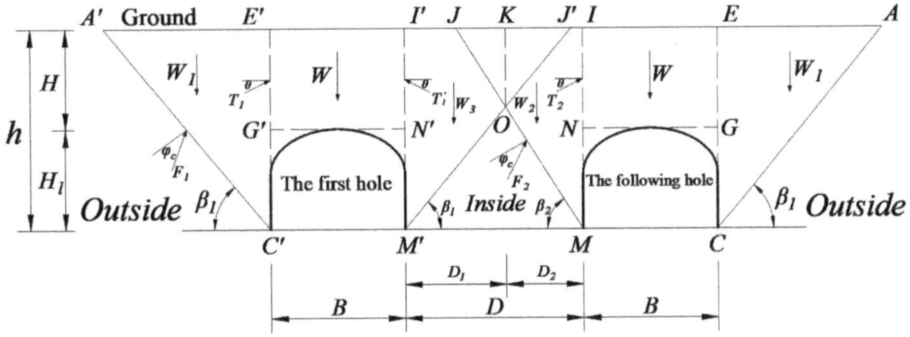

Figure 8. Theoretical sliding failure mode and calculating load schematic diagram.

3.2. Calculation of Theoretical Load

(1) Surrounding rock pressure of the first hole

Assume that the left hole is the first hole and excavated firstly. Since it is the same as an excavation of a single hole, the formula in the Chinese standard can be adopted to calculate the vertical pressure and the lateral pressure on the outside of the first hole. It is shown as follows:

The formula for calculating the lateral-pressure coefficient (λ_1) is:

$$\lambda_1 = \frac{\tan\beta_1 - \tan\varphi_c}{\tan\beta_1[1 + \tan\beta_1(\tan\varphi_c - \tan\theta) + \tan\varphi_c\tan\theta]} \tag{1}$$

$\tan\beta_1$ can be inferred from $\tan\varphi_c$ and $\tan\theta$:

$$\tan\beta_1 = \tan\varphi_c + \sqrt{\frac{(\tan^2\varphi_c + 1)\tan\varphi_c}{\tan\varphi_c - \tan\theta}} \tag{2}$$

The lateral pressure of the first hole (e_i) is:

$$e_i = \lambda_1\gamma h_i \tag{3}$$

where h_i represents the depth of the calculation position and γ means the unit weight of the rock.

Hence when the first hole is excavated solely, the vertical pressure on the vault of the hole (q) is:

$$q = \frac{Q}{B} = \gamma H(1 - \frac{H}{B}\lambda_1\tan\theta) \tag{4}$$

where B is the span of the hole, Q is the total vertical load on the hole and H is the depth of the vault of the hole.

(2) Surrounding rock pressure of the following hole

When it comes to the right hole (the following hole), the lateral pressure outside of the following hole is the same as the first hole. For calculating the lateral pressure inside of the following hole, a force analysis is carried out on the rock mass *IKOM*, as shown in Figure 9.

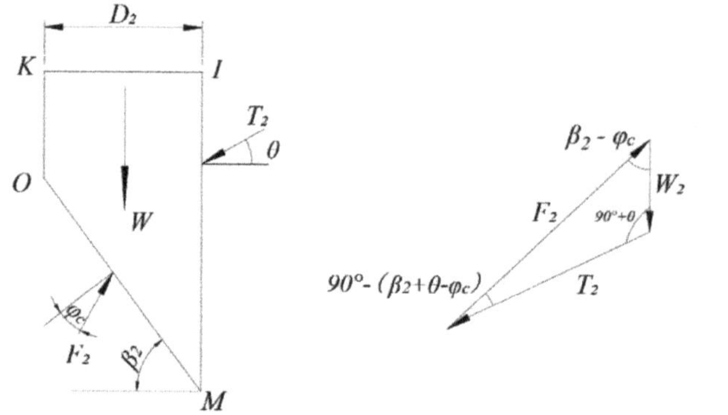

Figure 9. The force analysis of rock mass *IKOM*.

According to the basic mechanics and assumptions above, the force T_2 can be solved as:

$$T_2 = \frac{\sin(\beta_2 - \varphi_c)W_2}{\sin[90^\circ - (\beta_2 + \theta - \varphi_c)]} \tag{5}$$

where W_2 denotes the weight of the rock mass *IKOM*:

$$W_2 = \frac{1}{2}\gamma D_2(2h - D_2\tan\beta_2) \tag{6}$$

Seeing that the interval between the twin tunnels (D) is generally small enough, let $D_2 = D/2$, approximately. With the Equation (6), the Equation (5) can be transformed into:

$$T_2 = \frac{1}{2}\gamma h^2 \frac{\lambda_2}{\cos\theta} \tag{7}$$

In the Equation (7), λ_2 is defined as the inside lateral-pressure coefficient of the following hole:

$$\lambda_2 = \frac{\frac{D}{h}\left(1 - \frac{D}{4h}\tan\beta_2\right)(\tan\beta_2 - \tan\varphi_c)}{1 + \tan\beta_2(\tan\varphi_c - \tan\theta) + \tan]\varphi_c \tan\theta} \tag{8}$$

To make T_2 maximum, let $\frac{d\lambda_2}{d(\tan\beta_2)} = 0$, therefore

$$\tan\beta_2 = \sqrt{\frac{\tan^2\varphi_c + 1}{\tan\varphi_c - \tan\theta}\left[\frac{1}{\tan(\varphi_c - \theta)} + \frac{4h}{D}\right]} - \frac{1}{\tan(\varphi_c - \theta)} \tag{9}$$

Then, the lateral pressure of the following hole (e_2) is:

$$e_{2i} = \lambda_2 \gamma h_i \tag{10}$$

It is mentioned above that q is defined as the vertical pressure on the vault, whereas the vertical pressure on the vault actually varies from inside to outside. Suppose that the vertical pressure was changing linearly inside and outside of the hole (define the vertical pressure outside is q_1 and the other is q_2, q_1, and q_2 can be expressed as:

$$q_1 = \gamma H\left(1 - \frac{H}{B}\lambda_1 \tan\theta\right) \tag{11}$$

$$q_2 = \gamma H\left(1 - \frac{H}{B}\lambda_2 \tan\theta\right) \tag{12}$$

When the following hole is being excavated, the lateral pressure on the outside of the first hole can be regarded as a static value, approximately. Then, for calculating the lateral pressure on the inside of the first hole, a force analysis is carried out on the rock mass $I'KOM'$. As well as T_2 (Equation (5)), $T_2{}'$ can be solved as:

$$T_2{}' = \frac{1}{2}\gamma h^2 \frac{\lambda_2{}'}{\cos\theta} \tag{13}$$

Similarly, $\lambda_2{}'$ is defined as the lateral-pressure coefficient on the inside of the first hole when the following hole is being excavated:

$$\lambda_2{}' = \frac{\frac{D}{h}\left(1 - \frac{D}{4h}\tan\beta_1\right)(\tan\beta_1 - \tan\varphi_c)}{1 + \tan\beta_1(\tan\varphi_c - \tan\theta) + \tan\varphi_c \tan\theta} \tag{14}$$

$q_1, q_2{}'$ are defined as the vertical pressure on the outside and inside of the first hole, respectively. They can be written in the form:

$$q_1 = \gamma H\left(1 - \frac{H}{B}\lambda_1 \tan\theta\right) \tag{15}$$

$$q_2{}' = \gamma H\left(1 - \frac{H}{B}\lambda_2{}' \tan\theta\right) \tag{16}$$

Thus, the theoretical formulas for calculating the vertical pressure and lateral pressure of twin tunnels are deduced completely. Next, put all the parameters of Daling Tunnel into

Equations (1)–(16), the total surrounding rock pressure (theoretical load) was gained, as shown in Figure 3.

4. Results and Discussion

Generally, the surrounding rock pressure can be mainly divided into three parts: The first part of it is relaxing or transferring to the advance support and the surrounding rock before constructing the primary support. The second and third parts are supported by the primary support (share rate of the primary support) and the secondary lining (share rate of the secondary lining), respectively. By the result of simulation, calculation, and monitoring the share rate of these three parts can be identified.

4.1. Share Rate of the Primary Support

10%, 20%, 30%, 40% ... of the theoretical loads were put on the model gradually to match the measured displacement. When proportion of the theoretical load is 10%, 20%, 30%, 40%, etc., the vault settlement of the simulation model is 7.31 mm, 12.56 mm, 17.8 mm, 23.1 mm, respectively, which indicates that the vault settlement is growing as the load proportion is rising. Meanwhile, the measured vault settlement is shown in Figure 10a, which stabilized at 20.3 mm finally. In view of the lagging of the installation of the measuring device [36], the vault settlement of the model under 40% of the theoretical load (shown in Figure 10b) in accordance with the result of measurement. Consequently, share rate of the primary support is 40%.

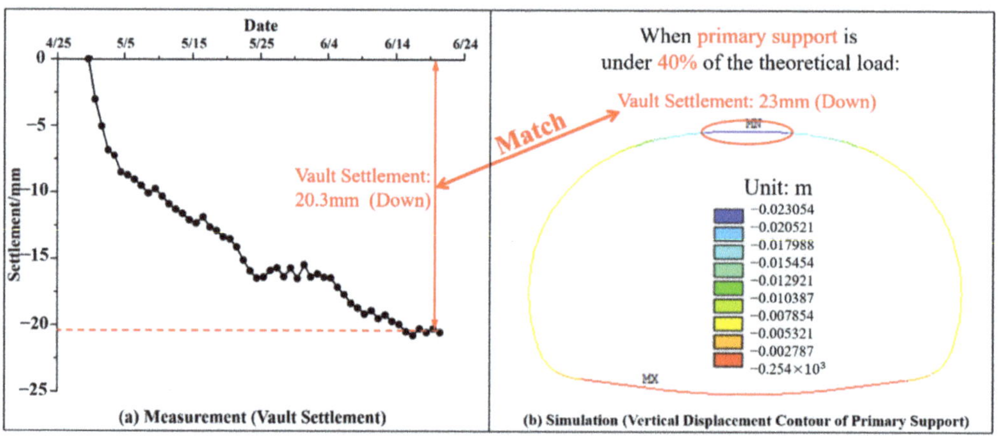

Figure 10. Measured and simulated vault settlement. When primary support is under 40% of the theoretical load, the result of simulation is in accord with the measurement.

4.2. Share Rate of the Secondary Lining

Further, to find the share rate of the secondary lining, the internal forces of the simulation model were compared with the actual structure. It was found that after stepwise loading, when the load ratio was 40%, internal forces of the simulation model were in good agreement with the practical measurement. Internal forces (axial force and moment) of the simulation model under 40% of the theoretical load is shown in Figure 11a,b.

The measured concrete stress of the inside and outside secondary lining is shown in Figure 12.

According to the theory of mechanics of materials, the actual internal forces can be obtained by the measured stress with the formulas below:

$$N = A \times (\sigma_{\text{inside}} + \sigma_{\text{outside}}) \tag{17}$$

$$M = \frac{I \times (\sigma_{\text{inside}} - \sigma_{\text{outside}})}{y} \tag{18}$$

where A (m^2) is the area of the cross section, I (m^4) is the moment of inertia of the section, y (m) is the distance between the neutral axis, and the edge, σ_{inside} (MPa) is the stress of the concrete at the side of the inner section and σ_{outside} (MPa) is the stress of the concrete at the side of surrounding rock.

Internal force was gained after putting the related parameters and stress into the formulas. Then, according to the Chinese standard [1] based on the internal force and the capacity of the concrete structure the safety factors were calculated. It is shown in Table 3 and Figure 13.

$$\left(Safetyfactor = \frac{capacity\,of\,the\,concrete\,structure}{stress\,of\,the\,structure}\right)$$

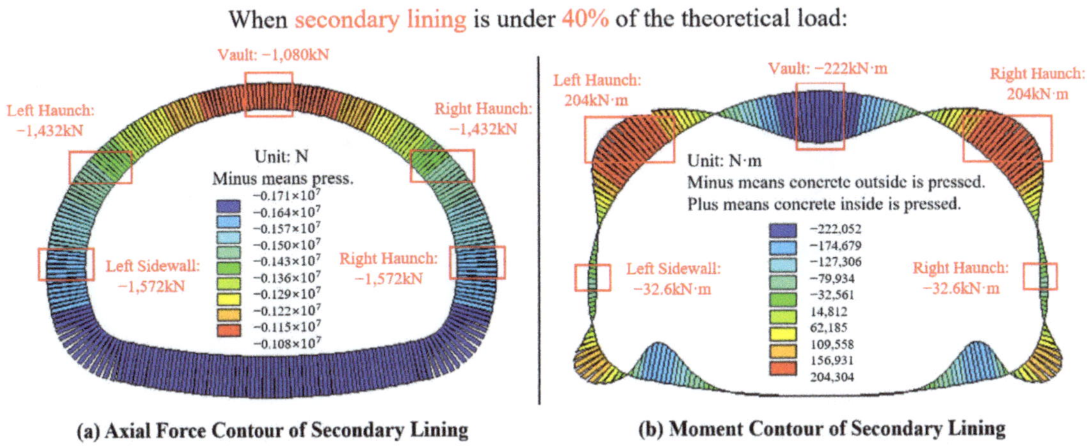

Figure 11. Moment (N·m) and axial-force (N) diagram of the secondary lining in simulation.

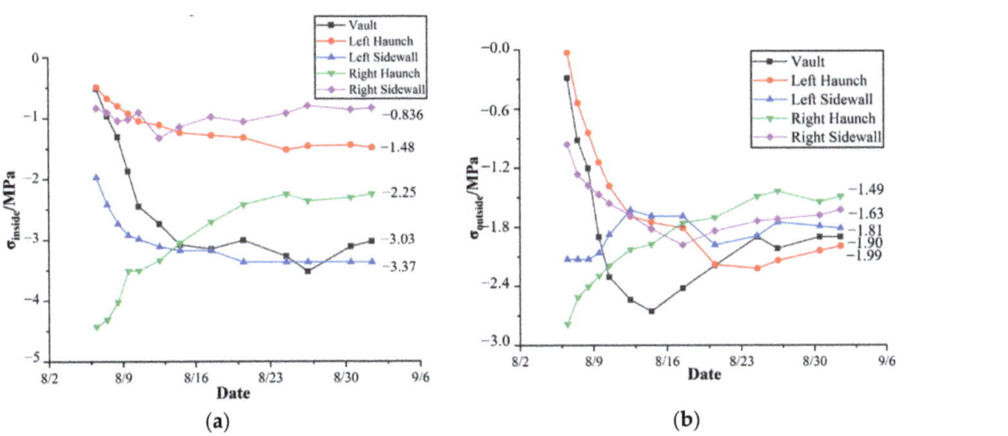

Figure 12. Measured concrete stress of the secondary lining. It went stable after about a month. (**a**) inside secondary lining; (**b**) outside secondary lining.

Table 3. The internal forces and safety factors of the secondary lining.

Item	Position	Axis Force /kN	Moment /kN·m	Eccentricity /mm	Attribute	Safety Factor
Simulation	Vault	−1080	−222	493	All is small eccentricity	9.47
	Left haunch	−1432	204	430		8.19
	Left sidewall	−1572	−32.6	308		10.41
	Right haunch	−1432	204	430		8.19
	Right sidewall	−1572	−32.6	308		10.41
Measurement	Vault	−1723	−53	318		7.85
	Left haunch	−1216	24	307		11.54
	Left sidewall	−781	71	378		14.0
	Right haunch	−1348	−36	314		10.17
	Right sidewall	−862	38	332		15.00

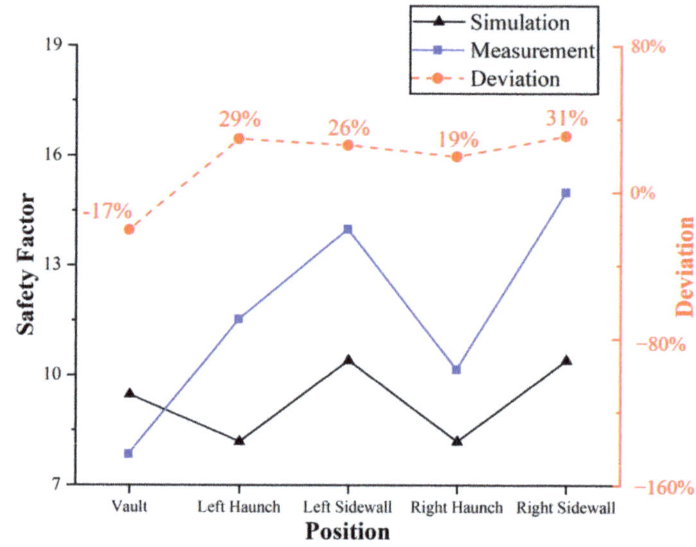

Figure 13. Safety factors of simulation and measurement. The deviation is about 25% averagely and mainly attributed to the efficiency of the rebar in secondary lining.

According to Table 3 and Figure 13, the safety factor of the simulation model is slightly different from the measurement. Generally, the safety factor of the simulation model is about 25% smaller than the measurement. This phenomenon is mainly because the efficiency of the rebar in the secondary lining is not considered during the calculation. Therefore, it is feasible to believe that the share rate of the secondary lining is 40%.

4.3. *Total Pressure for Verification*

The pressure cells were arranged at the points A, B, and C (in Figure 5) outside of the primary support. They can measure the total pressure of the primary support and the secondary lining. Here, the result of pressure cells is compared with 40% + 40% = 80% of the theoretical load, and it is shown in Figure 14.

It can be seen from the Figure 14a that the pressures at point A, B, and C were surging rapidly after the excavation, then they stabilized in approximately one month. Furthermore, the pressure at vault (Point A) was the largest, haunch (Point B) was the second and sidewall (Point C) was the smallest.

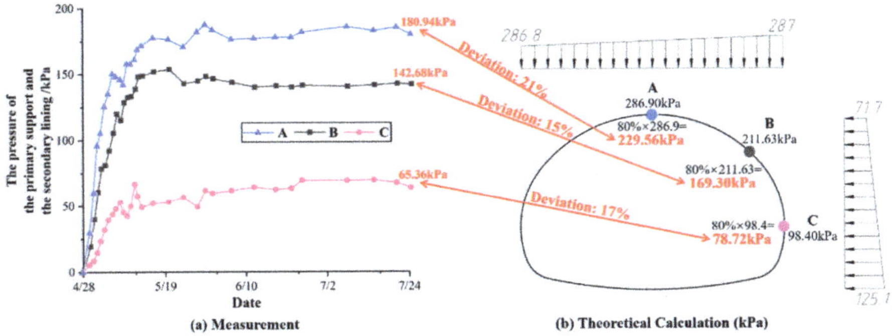

Figure 14. Total pressure of the primary support and the secondary lining. There is about 20% deviation at each point and it is believed to be born mainly by advance support and the surrounding rock itself.

4.4. Discussion

According to the content of Sections 4.1 and 4.2, the share rate of the primary support and the secondary lining are both reckoned to be 40%. The other 20% of the surrounding rock pressure is taken by multiple factors (mainly the advance support and the surrounding rock itself).

In other words, the aggregated rock pressure borne by primary support and secondary lining is 80% (of the theoretical load). Combined with the content of Section 4.3, the comparison of the calculation and the measurement is shown in Table 4 and Figure 14.

Table 4. Comparison of the calculation and measurement.

Measured Point	Theoretical Load /kPa	80% of the Theoretical Load /kPa	Measured Pressure /kPa	Deviation /%
A	286.90	229.56	180.94	21.18
B	211.63	169.30	142.68	15.73
C	98.40	78.72	65.36	16.97

From Table 4, measured pressure, which the rock pressure that the primary support and secondary lining exactly bear, is less than 80% of the theoretical load according to the measurement. The main reason for this deviation is that before the installation of the pressure cells and the construction of the primary support, part of the rock pressure was relaxed and cannot be measured. Usually, this part of the rock pressure was supported by the surrounding rock mass itself. So, in consideration of safety, the conclusion that total share rate of the primary support and the secondary lining is 80% is valid for design. The correctness of theoretical calculation of surrounding rock pressure is also verified by Table 4 and Figure 14 as the deviation is acceptable.

In summary, based on the analysis above, the share rates of the primary support and the secondary lining are both 40% in Daling Tunnel (the super-large span twin tunnel), and other 20% of pressure is mainly born by the advance support and the surrounding rock itself. Although the results of theoretical calculation and the measurement do not match perfectly, the overall deviation is small enough, which basically meets the requirements of design and application in engineering.

Meanwhile, during the whole process, there are several assumptions and limitation, for example:

(1) The influence of the construction's sequence of the first hole and following hole was not considered in this research, and the way of excavation was assumed as full section excavation.

(2) This paper supposed that the stratum was consistent and the surrounding rock mass was ideal.
(3) The surface of earth was presumed to be horizontal, and so on.

These assumptions and limitation are supposed to be concerned when the conclusion is applied to the practical projects. In addition, the share rate is affected by the strength and stiffness of the composite lining as well as the method and time of excavation and construction. Thus, the share rate of the twin tunnels with super-large span in various conditions still needs to be researched systematically.

5. Conclusions

Based on the theory of stratigraphic slip failure mode and comparison of the simulation and the measurement, the share rate of twin tunnels with super-large span and small interval are studied in this paper. Furthermore, the analysis was verified and supplemented by the outcome of monitoring and measuring. Finally, the conclusions are drawn as follows:

(1) The formulas of calculating surrounding rock pressure of normal tunnels (usually including two or three lanes) in Chinese standard could be applied to the super-large span twin tunnels (including four or more lanes). The method was verified by field measurement.
(2) The method of researching the share rate of the primary support and the secondary lining of the tunnel was proposed. The result demonstrates that the share rate of the primary support and the secondary lining of Daling Tunnel are both 40%. Usually for twin tunnels with super-large span in cities, not only the primary support but also the secondary lining should be strong enough to ensure the design is safe and reliable. In addition, this conclusion is conservative when compared to the measurement, which is practical for application in engineering.
(3) The result of the theoretical calculation and numerical simulation was compared with the measurement to evaluate the research methodology and achievements. These two results match properly, which verify the correctness of this study. Hence the conclusions and the method of this research can make some reference to the design, construction, and maintenance of super-large span twin tunnels.

Since this research is specific for twin tunnels with four lanes (each hole) in cities, this paper can make great contributions to urbanization, infrastructure construction, and even standards drafting. Moreover, this research is more focused on the first hole and does not consider the influence of construction sequence. Therefore, the conclusions drawn in this paper are more appropriate for the first hole.

Author Contributions: Conceptualization: K.S.; data curation: Y.Z. and Y.H.; formal analysis: X.Z.; investigation: X.Z., Y.Z., Y.H. and H.Z.; methodology: K.S.; software: X.Z., Y.H. and H.Z.; writing—original draft: K.S. and H.Z.; writing—review and editing: X.Z. and Y.Z. All authors have read and agreed to the published version of the manuscript.

Funding: This research was funded by the National Natural Science Foundation of China, grant number: 52178396.

Institutional Review Board Statement: Not applicable.

Informed Consent Statement: Not applicable.

Data Availability Statement: The data of simulation and monitoring in the article is not freely available due to legal concerns and commercial confidentiality. Nevertheless, all the concepts and procedures are explained in the presented research and parts of the research may be available upon request.

Acknowledgments: The authors of the present work feel grateful and would like to thank Shandong Expressway Jitai Intercity Highway Co., Ltd., Shandong Expressway Jilai Intercity Highway Co., Ltd., National Natural Science Foundation of China, Southwest Jiaotong University, for providing facilities, fund and supports in measurement and monitoring.

Conflicts of Interest: The authors declare that they have no conflict of interest.

References

1. China Railway Eryuan Engineering Group Co., Ltd. Specifications for Design of Highway Tunnels. In *Civil Engineering*; China Transportation Press: Beijing, China, 2018; JTG 3370.1-2018.
2. Feng, J.M.; Yan, C.W.; Ye, L.; Ding, X.Q.; Zhang, J.R.; Li, Z.L. Evaluation of installation timing of initial ground support for large-span tunnel in hard rock. *Tunn. Undergr. Space Technol.* **2019**, *93*, 103087. [CrossRef]
3. Hou, F.J.; Li, S.C.; Li, W.J.; Li, X.Z.; Han, X.M. Mechanical Analysis of Large-span Tunnel Construction in Urban Seaside. *J. Coast. Res.* **2019**, *94*, 255–259. [CrossRef]
4. Hou, F.J.; Han, X.M.; Li, S.C.; Li, W.J. Optimization Research of Construction Methods and Parameters of Large-span City Coastal Tunnel in Horizontal Layered Rockmass. *J. Coast. Res.* **2019**, *94*, 232–236. [CrossRef]
5. Li, A.; Zhang, D.L.; Fang, Q.; Luo, J.W.; Cao, L.Q.; Sun, Z.Y. Safety Distance of Shotcrete Subjected to Blasting Vibration in Large-Span High-Speed Railway Tunnels. *Shock. Vib.* **2019**, *2019*, 2429713. [CrossRef]
6. Li, R.; Zhang, D.L.; Fang, Q.; Li, A.; Hong, X.F.; Ma, X.B. Geotechnical monitoring and safety assessment of large-span triple tunnels using drilling and blasting method. *J. Vibro Eng.* **2019**, *21*, 1373–1387.
7. Sun, S.C.; Rong, C.X.; Wang, H.L.; Cui, L.Z.; Shi, X. The Ground Settlement and the Existing Pipeline Response Induced by the Nonsynchronous Construction of a Twin-Tunnel. *Adv. Civ. Eng.* **2021**, *2021*, 8815304. [CrossRef]
8. Liu, C.; Li, S.C.; Zhou, Z.Q.; Li, L.P.; Shi, S.S.; Chen, Y.X. Numerical Analysis of Surrounding Rock Stability in Super-Large Section Tunnel Based on Hydro-Mechanical Coupling Model. *Geotech. Geol. Eng.* **2019**, *37*, 1297–1310. [CrossRef]
9. Chortis, F.; Kavvadas, M. Three-Dimensional Numerical Investigation of the Interaction Between Twin Tunnels. *Geotech. Geol. Eng.* **2021**, *39*, 5559–5585. [CrossRef]
10. Vinod, M.; Khabbaz, H. Comparison of rectangular and circular bored twin tunnels in weak ground. *Undergr. Space* **2019**, *4*, 328–339. [CrossRef]
11. Zhang, D.L.; Chen, L.P.; Fang, Q.; Song, R.G. Research and application on central rock wall dike stability of small interval tunnel. *J. Beijing Jiaotong Univ.* **2016**, *40*, 1–11.
12. Boon, C.W.; Neo, C.W.; Ng, D.C.C.; Ong, V.C.W. Discontinuum analyses of openings constructed with side drift and limited rock cover. *J. Zhejiang Univ.-SCIENCE A* **2018**, *19*, 255–265. [CrossRef]
13. Ghorbani, H.; Ajalloeian, R. A method for selection of optimum distance between twin tunnels under static and pseudo-static conditions, case study: Pooneh tunnel. *Geotech. Geol. Eng.* **2019**, *37*, 4435–4446. [CrossRef]
14. Nematollahi, M.; Molladavoodi, H.; Dias, D. Three-dimensional numerical simulation of the Shiraz subway second line - influence of the segmental joints geometry and of the lagging distance between twin tunnels' faces. *Eur. J. Environ. Civ. Eng.* **2020**, *24*, 1606–1622. [CrossRef]
15. He, C.; Zhou, S.H.; Di, H.G.; Yang, X.W. Effect of Dynamic Interaction of Two Neighboring Tunnels on Vibrations from Underground Railways in the Saturated Soil. *KSCE J. Civ. Eng.* **2019**, *23*, 4651–4661. [CrossRef]
16. Khabbaz, H.; Gibson, R.; Fatahi, B. Effect of constructing twin tunnels under a building supported by pile foundations in the Sydney central business district. *Undergr. Space* **2019**, *4*, 261–276. [CrossRef]
17. Zhu, J.L.; Zhu, D.Y. Deformation of Pipelines Induced by the Construction of Underlying Twin-Tunnel. *Teh. Vjesn.-Tech. Gaz.* **2020**, *27*, 1311–1315.
18. Yang, Z.H.; Li, S.S.; Yang, D.; Zhou, S.; Zhou, Q. Stability of Twin Shallow Tunnels in Unsaturated Soils Considering Vertical Steady State Flow. *Geotech. Geol. Eng.* **2019**, *37*, 4603–4612. [CrossRef]
19. Golshani, A.; Varnusfaderani, M.G. Innovative design modification during construction of a twin tunnel using real-time field data. *Transp. Geotech.* **2019**, *20*, 100254. [CrossRef]
20. Komu, M.P.; Guney, U.; Kilickaya, T.E.; Gokceoglu, C. Using 3D Numerical Analysis for the Assessment of Tunnel-Landslide Relationship: Bahce-Nurdag Tunnel South of Turkey. *Geotech. Geol. Eng.* **2020**, *38*, 1237–1254. [CrossRef]
21. Bai, B.; Yang, G.C.; Li, T.; Yang, G.S. A thermodynamic constitutive model with temperature effect based on particle rearrangement for geomaterials. *Mech. Mater.* **2019**, *139*, 103180. [CrossRef]
22. Bai, B.; Zhou, R.; Cai, G.Q.; Hu, W.; Yang, G.C. Coupled thermo-hydro-mechanical mechanism in view of the soil particle rearrangement of granular thermodynamics. *Comput. Geotech.* **2021**, *137*, 104272. [CrossRef]
23. Bai, B.; Wang, Y.; Rao, D.Y.; Bai, F. The effective thermal conductivity of unsaturated porous media deduced by pore-scale SPH simulation. *Front. Earth Sci.* **2022**, *10*, 943853. [CrossRef]
24. Sun, K.G.; Hong, Y.Q.; Xu, W.P.; Hou, Z.H.; Liu, X.; Yu, M.Z.; Yuan, Z.Y. Analysis and prediction of mechanical characteristics of corrugated plate as primary support in tunnels. *Tunn. Undergr. Space Technol.* **2021**, *111*, 103845. [CrossRef]
25. Sun, K.G.; Hong, Y.Q.; Xu, W.P.; Liu, H.; Zhen, Y.Z.; Qin, J.H. Analysis and prediction of the mechanical behavior of corrugated plate as primary support in tunnels with elastoplastic constitution. *Tunn. Undergr. Space Technol.* **2022**, *124*, 104451. [CrossRef]
26. Aydin, E.; Dutkiewicz, M.; Öztürk, B.; Sonmez, M. Optimization of elastic spring supports for cantilever beams. *Struct. Multidiscip. Optim.* **2020**, *62*, 55–81. [CrossRef]
27. Wang, Y.H.; Nguyen, N.H.T. The effects of rock-infill interfacial properties on the compressive damage behaviour of flawed rocks: Results from a DEM study. *Theor. Appl. Fract. Mech.* **2022**, *117*, 103166. [CrossRef]

28. Wang, Y.H.; Nguyen, N.H.T.; Zhao, L.H. Micromechanical study on hard rock strainburst using the discrete element method. *Tunn. Undergr. Space Technol.* **2021**, *109*, 103793. [CrossRef]
29. Li, P.F.; Wang, F.; Fan, L.F.; Wang, H.D.; Ma, G.W. Analytical scrutiny of loosening pressure on deep twin-tunnels in rock formations. *Tunn. Undergr. Space Technol.* **2019**, *83*, 373–380. [CrossRef]
30. Guo, Z.H.; Liu, X.R.; Zhu, Z.Y. Elastic Solution for a Deep Twin Tunnel Stress Based on Complex Variable Theory and the Superposition Principle. *J. Eng. Res.* **2017**, *5*, 68–86.
31. Panji, M.; Kavandi, P. Investigating of Effective Parameters on Stress Behavior of Pressure Deep Twin Tunnels under Pressure. Turk Online J. *Des. Art Commun.* **2016**, *6*, 1184–1189.
32. Zhong, Z.L.; Liu, X.R.; Liu, Y.X.; Zhang, J. Theoretical calculation and monitoring on rock pressure of shallow-buried bilateral bias neiborhood tunnel. *J. Chongqing Univ.* **2013**, *36*, 63–68.
33. Lyu, H.M.; Shen, S.L.; Zhou, A.N.; Chen, K.L. Calculation of pressure on the shallow-buried twin-tunnel in layered strata. *Tunn. Undergr. Space Technol.* **2020**, *103*, 103465. [CrossRef]
34. Kong, F.C.; Lu, D.C.; Du, X.L.; Shen, C.P. Elastic analytical solution of shallow tunnel owing to twin tunnelling based on a unified displacement function. *Appl. Math. Model.* **2019**, *68*, 422–442. [CrossRef]
35. Cui, S.S.; Wu, K.; Zhang, Q.J.; Yu, Y.L.; Wang, Y.J. Analysis on Construction Mechanical Mechanism of Highway Tunnels with Large Span and Small Spacing. *Geotech. Geol. Eng.* **2019**, *37*, 1627–1642. [CrossRef]
36. Qiu, J.L.; Liu, H.Q.; Lai, J.X.; Lai, H.P.; Chen, J.X.; Wang, K. Investigating the Long-Term Settlement of a Tunnel Built over Improved Loessial Foundation Soil Using Jet Grouting Technique. *J. Perform. Constr. Facil.* **2018**, *32*, 04018066. [CrossRef]

MDPI
St. Alban-Anlage 66
4052 Basel
Switzerland
www.mdpi.com

Applied Sciences Editorial Office
E-mail: applsci@mdpi.com
www.mdpi.com/journal/applsci

www.ingramcontent.com/pod-product-compliance
Lightning Source LLC
Chambersburg PA
CBHW061557210926
51850CB00009B/2135
* 9 7 8 3 7 2 5 8 0 0 4 2 1 *